T0048463

The European Research Council

The European Research Council

Thomas König

polity

Sophie Dvořák www.sophiedvorak.net

First published in 2017 by Polity Press

Polity Press
65 Bridge Street
Cambridge CB2 1UR, UK

Polity Press
350 Main Street
Malden, MA 02148, USA

ISBN-13: 978-0-7456-9124-4

A catalogue record for this book is available from the British Library.

Library of Congress Cataloging-in-Publication Data

Names: Konig, Thomas, 1976-
Title: The European Research Council / Thomas Konig.
Description: Cambridge : Polity, 2016. | Includes bibliographical references and index.
Identifiers: LCCN 2016025715 (print) | LCCN 2016026583 (ebook) | ISBN 9780745691244 (hardback) | ISBN 9780745691275 (Mobi) | ISBN 9780745691282 (Epub)
Subjects: LCSH: Research--European Union countries. | Research--Political aspects--Europe. | European Research Council.
Classification: LCC Q180.E85 K66 2016 (print) | LCC Q180.E85 (ebook) | DDC 507.2/04--dc23
LC record available at https://lccn.loc.gov/2016025715

Typeset in 10.5 on 12 pt Sabon by
Servis Filmsetting Ltd, Stockport, Cheshire
Printed and bound in Great Britain by Clays Ltd, St Ives PLC

For further information on Polity, visit our website: politybooks.com

CONTENTS

FIGURES

ABBREVIATIONS

AdG	Advanced Grant
AEI	Archive of European Integration (University of Pittsburgh, USA)
ALLEA	All European Academies
CAP	Common Agricultural Policy
CERN	European Organization for Nuclear Research
CNERP	Committee for a New European Research Policy
CNRS	Centre national de la recherche scientifique (France)
CoG	Consolidator Grant
COIME	Standing Committee on Conflict of Interest
COST	European Coordination in Science and Technology
DFG	Deutsche Forschungsgemeinschaft (Germany)
DG	Directorate General
DIS	Dedicated Implementation Structure
EA	Executive Agency
EC	European Commission
ECU	European Currency Unit
ECOFIN	Economic and Financial Affairs Council (Council of the European Union)
EARTO	European Association of Research and Technology Organizations
ECA	European Court of Auditors
EIRMA	European Industrial Research Management Association
EIT	European Institute of Innovation and Technology
ELSF	European Life Sciences Forum
ELSO	European Life Scientist Organization
EMBL	European Molecular Biology Laboratory
EMBO	European Molecular Biology Organization

ERA	European Research Area
ERC	European Research Council
ERCEA	European Research Council Executive Agency
ERCEG	European Research Council Expert Group
ESF	European Science Foundation
EU	European Union
EURAB	European Research Advisory Board
EURYI	European Young Investigator Awards
FEBS	Federation of European Biochemical Societies
FP	Framework Programme for Research and Technological Development
FTE	Full-time equivalents
GERD	Gross domestic expenditure on research and development
ISE	Initiative for Science in Europe
ITRE	Industry, Research and Energy (Committee of the European Parliament)
JRC	Joint Research Centre (European Commission)
LS	Life Sciences (ERC domain of research fields)
MFF	Multiannual Financial Framework
MPG	Max Planck Gesellschaft (Germany)
NEST	New and Emerging Science and Technology
NIH	National Institutes of Health (USA)
NSF	National Science Foundation (USA)
OECD	Organization for Economic Co-operation and Development
PE	Physics and Engineering (ERC domain of research fields)
PI	Principal Investigator
R&D	Research and Development
REA	Research Executive Agency
RISE	Research, Innovation and Science Policy Experts (European Commission expert group)
RJ	Riksbankens Jubileumsfond (Sweden)
ScC	Scientific Council (ERC)
SH	Social Sciences and Humanities (ERC domain of research fields)
S&T	Science and Technology
SPRU	Science Policy Research Unit (Sussex University, UK)
StG	Starting Grant
SyG	Synergy Grant

ABBREVIATIONS

UNESCO	United Nations Educational, Scientific and Cultural Organization
WEF	World Economic Forum
WP	World Programme

ACKNOWLEDGEMENTS

I am indebted to many people. To Michael Stampfer, for setting me off on the ERC route; to Helga Nowotny, for triggering my initial interest in the ERC and supporting this project without ever intervening in my interpretation of events, and to Barbara Blatterer, for the time spent together in the office in Vienna. Michèle Lamont gave me ideas, and Johannes Pollak provided me with a roof over my head. I am grateful to all the people who granted me the time to interview them, or allowed me access to their personal archive, or both (see the Appendices). Most importantly, Dan Brändström, William Cannell, Mogens Jensted-Flensted, Luc van Dyck, Anastasia Andrikopoulou, Alejandro Martin-Hobdey, Jens Degett, and Pavel Exner helped me to fill many gaps in the period up until 2007. Special thanks go to Michael Kwakkelstein and Tjarda Vermeijden from the Nederlands Interuniversitair Kunsthistorisch Instituut in Florence, Italy. At several of its meetings, a group of aspiring scholars in the 'European Research Area' network kindly commented on my unfinished thoughts on European research funding, peer review, and autonomy; special thanks also to Meng-Hsuan Chou, Tim Flink and Mitchell Young. Four anonymous reviewers gave crucial advice on an earlier draft of the manuscript. Several colleagues and scholars commented on various parts of this manuscript; those not yet mentioned are Ben Turner, Jerzy Langer, Karl-Ulrich Mayer, Michael Solberg, and Christian Fleck. John Thompson, George Owers, and the team at Polity have been extremely gracious and tolerant with me when one deadline after another was missed.

Writing a book is one of the most intellectually rewarding endeavours I can think of. It has so many layers of complexity – the facts, the archives, the structure, the fabric of the arguments, the narrative, and the tone, to mention just the most obvious ones. To write it, the Swedish Riksbankens Jubileumsfond provided me with a generous grant (number INT:13–1360–1). I benefited from a research scholarship to Harvard University's Center for European Studies between April and August 2014, sponsored by the Austrian-American Educational Commission (Fulbright Austria), and a writing retreat in the South of France, in the summer of 2015: *La Fondation des Treilles, créée par Anne Gruner Schlumberger, a notamment pour vocation d'ouvrir et de nourrir le dialogue entre les sciences et les arts afin de faire progresser la création et la recherche contemporaines. Elle accueille également des chercheurs et des écrivains dans le domaine des Treilles (Var)*. There have been other havens for writing uninterruptedly, too: the Institute for Advanced Studies in Vienna (IHS) generously offered me an office from October 2014 onwards. Parts of this book were also written during vacation time in Graz, in the attic of my parents' house, and in Ottakring, Vienna, in the apartment of my Best Man, Christian Würth.

Writing this book was also one of the most solitary activities I have ever embarked upon. Even while watching our toddler son learning to walk, I caught myself thinking of the most appropriate translation of a peculiar word, such as 'Bezeichnung', in English. I have been negligent of many friendships over the past years. More than anything, I owe this book to Melissa. She was an unflagging source of emotional support in bringing this project to a conclusion, while all the time our priorities as a family were shifting dramatically: our son Anton was born when I had first begun thinking about this project, and our daughter Lucy was born just as I was in the final stretch of writing it down. To them, I dedicate this book.

INTRODUCTION

> Patience, tact, good manners, and a fresh but not innocent eye, are all desirable characteristics for a field worker.[1]

Place Rogier, or Rogierplein (depending on which of the two official Belgian languages one prefers) constitutes the southeastern corner of the smallest of Brussels' municipalities, Saint-Josse-ten-Noode or Sint-Joost-ten-Node. It is an industrious place: Rue Neuve/Nieuwstraat, one of the city's major commercial streets, terminates here; the Boulevard du Jardin Botanique/Kruidtuinlaan passes through it, as part of the inner ring road surrounding the city centre; underneath, metro lines 2 and 6 cross several tram lines, complemented by bus routes at ground level. People also flock in from the Gare du Nord/Noordstation, which is just one block to the north, surrounded by a large red-light district. The site's most significant landmark for the past years (though allegedly temporarily) has been a hole in its centre, dug and maintained with the intention of modernizing the underground infrastructure. Several hotels, a large bank tower, restaurants and food chains, as well as two shopping plazas complete the setting.

Sitting uncomfortably at its edge is also a newly erected building complex called Covent Garden (no French or Flemish name available), which consists of two high-rise buildings and is part of the Northern Quarter (Quartier Nord/Nordwijk), Brussels' informal business district with circa 40,000 employees. The European Research Council Executive Agency (ERCEA) has been lodged here since mid–2009. Entering the building through its revolving doors, a visitor to the agency is instantly lulled into an artificial quietude, which is even more striking after traversing the crowded and noisy square outside. The glazed atrium features trees several metres high, 'comfort zones' with white leather ottomans for chatting and relaxing, and (rather inconveniently) a cobblestone pavement.

Between 2010 and 2013, I have been a regular visitor to the ERCEA; according to my own accounts, I have spent about forty workdays a year there. I did so in my capacity as the scientific adviser to the President of the ERC. The President, Helga Nowotny, was also the chair of the ERC Scientific Council. If the ERCEA was the executive arm of the European Research Council, the Scientific Council was its strategy-devising head. Contrary to what one may think, however, there were hardly any formal links between these two entities. Nowotny had an office at the ERCEA premises, but her home base was a smallish research funding organization in Vienna, which was also my employer. Due to this arrangement, my way in at Covent Garden used to be barred by security gates. Following European Commission standard security protocol, I had to give my passport number, provide a contact name at the agency, and wait for this person to pick me up. It was a repeated annoyance not so much for me but for the particular person in the ERCEA Secretariat I phoned five minutes in advance to say that I would be arriving and to request that she or he come down.

After roughly the first half of my active involvement with the ERC, I was given a magnetic badge with which I could pass through the security gate and move freely through the entire premises – well, probably not the entire premises, but those parts I had to visit. The issue of a badge was decided by the ERCEA director, then (as now) Pablo Amor, a long-standing European Commission functionary, and it required – so I was told – the bending of a certain security protocol; I never learned which, nor what the bending consisted of. Amor must have come to the conclusion that I was sufficiently innocuous to be granted this extraordinary privilege. To me, the badge signified that, although nobody had a clue what my advisory role actually entailed, I had become sufficiently trustworthy. More generally, Amor's decision was a symbol of the growing mutual confidence between the ERCEA management and the ERC Scientific Council leadership, and Helga Nowotny in particular. This was in early 2012, and it was around this time that I decided to write a book on the ERC.

The ERCEA is situated in the five top floors of Covent Garden's tower B, the higher of the two high-rise buildings; its nerve centre is the 24th floor, which is also where I was usually heading when admitted downstairs. Most of the large meeting rooms for evaluation panel meetings are located here, as are the offices of the ERC leadership; both are equipped with floor-to-roof windows, providing a spectacular view over the sprawling city. The floor also hosts a huge Nespresso coffee machine, after many complaints about the allegedly appalling coffee served by a catering company contracted by the European Commission; it is an exclusive privilege in the entire building, accounted for by the relative autonomy of an Executive Agency and its Director (as was the approval of my badge). On the platform of the six elevator shafts on that floor hangs a sign with the ERC logo on top, oddly dated in Roman numerals ('ERCEA MMXII D.C.'), and a quotation in capital letters:

> The European Research Council Executive Agency is dedicated to selecting and funding the excellent ideas that have not happened yet and the scientists that are dreaming them up.

'Selecting' and 'funding' describe in a nutshell everything that a funding instrument like the ERC is tasked to fulfil. Yet, in stark contrast to the grandiose rhetoric of the quotation it carries, the sign is printed on a cheap looking woven fabric; I never found out whether this was an unconscious result of the cost-cutting mentality that drives a European Commission executive agency, or whether it was cleverly playing with this assumption. In any case, the palpable discrepancy nicely brings to the fore that, while they constitute two sides of the same overall purpose, here are two different principles that have to be brought together: applying scientific values to the selection of suitable proposals to be carried out, and ensuring the expedient (contractually legal) usage of public funds that are spent on this activity. In retrospect, whatever I was doing at the ERC, it all revolved, in one way or other, around one of those two sets of procedures – either working along their rules or attempting to improve them.

My time at the ERC was basically divided into four routines. The first routine was the annual submission of a 'proposal' to the ERCEA in order to secure funding for the office of the Scientific Council chair in Vienna and those of her deputies. It meant that I had to write up a 'project' outlining our role in supporting the Scientific Council leadership trio, attempting to come as close as possible to a pre-fixed amount of money. Once the proposal had been submitted and 'evaluated' by my colleagues in the ERC Executive Agency, I had to 'negotiate' with them. It was a necessarily awkward exercise, not only because the 'project' consisted of deliverables that were simply taken from the Scientific Council's annual schedule of meetings (I believe no one ever really checked them anyway), but also in the way the formal process of negotiating the contract was detrimental to my daily interactions with the ERCEA staff, with whom I was discussing issues such as how to prepare a certain working group meeting of the Scientific Council, or how to pitch a public statement by the President about how great and well-functioning the ERC had become.[2] A conflict of interest? Certainly not!

The second routine was to help Nowotny put together the ERC panels in the area of social sciences and humanities, for which she was responsible. Usually, it started with my contacting confidants of Nowotny to ask them for names of potential panel members; that is, rising stars and ambitious newcomers in one of the numerous fields that were covered by this subject area, ranging from economics to philology, and from psychology to art history. Next, I searched for the CVs of those people (and others that I found similarly qualified for the position), then I prepared short profiles of each of them to be studied by Nowotny. It was her duty to judge potential panelists' reputations and their intellectual capacity; once she had approved a set of profiles, my colleagues at the ERCEA's Scientific Department and I could actually put together the panels. In doing so, we were concerned with additional practical things: all panelists should have a good command of English; furthermore, one panel should consist of evaluators as diverse as possible in terms of disciplinary expertise, gender, nationality, and possibly also age. Putting together panels was an on-going exercise, not only because the ERC funding machinery was expanding so quickly,[3] but also because the panels required constant refinement and replacement of panellists, who either dropped out or were dismissed because of a lack of commitment. Although this was time-consuming work, I enjoyed it considerably, as it gave me a first-hand overview of the newest trends in different fields – areas in which I had had little experience previously (linguistics! geography! philology!), their current state of affairs and recent ground-breaking publications.

The third routine was to spend time at Covent Garden. That may sound strange, but talking, listening, observing, making notes about all sorts of encounters, from panel meetings to informal meetings for coffee, were part of my mission to have an ear to the ground in the Executive Agency, to know about any difficulties with staff members, and to distribute informally new developments in the Scientific Council and its leadership. While I cannot estimate to what degree my efforts contributed to the overall goal, the reason for this routine was to build and maintain what I have later called the 'socio-organizational fabric' between the two entities in the ERC compound.[4]

The fourth routine was probably the most traditional that one may expect if one thinks of someone being an 'adviser'. It was to prepare Scientific Council meetings and to draft letters for the President. If I say 'to prepare', I basically mean to read, for, if I remember correctly, I was not really ever being asked to draft any of the documents used by the Scientific Council. I took some pride in occasionally pointing out a few potential flaws to Nowotny. Only when doing research on this book did I learn that, by the time I began this work for the Scientific Council, this was a left-over from earlier times, when those documents potentially contained grave misunderstandings between the two bodies, the ERCEA and the Scientific Council – misunderstandings on a scale I do not recall from the time I was with the ERC. I wrote draft letters and presentations for real, however, and it may have been from this subset of the routine that I got my first ideas about reflecting on the status of research funding more generally.

The ERC years were a great opportunity for me to learn something about European politics and the intricacies of the European integration project in general, and they also enhanced my practical knowledge of the relationship between science and policy in one of their most fruitful yet fragile areas of contact: money. The first real motivation to write a book on the ERC was when I realized that many of the anecdotes about the early days of the ERC that I had been told by Scientific Council members didn't quite add up. To be sure, the quips were different in content and style – some were told as *bon mots*, some as crucial, decisive moments; some anecdotes were belligerent, relating to a triumph, a victory, albeit occasionally a pyrrhic or tragic one; others spoke of an achievement, an extraordinary result through negotiations, or even a miracle; others again claimed an effect, a change of behaviour or an altered trajectory. What irritated me was that, when trying to put them together, there were inconsistencies in timing and in arguments. My initial idea was to clarify those inconsistencies and to sum up gently the history of the ERC in an easy-going mode. Nothing too academic, yet with some interesting twists.

With my ambition to tell the real story (and to tell it as a great story), the first real step towards writing this book seriously took off during the last few months of my time with the ERC, when I started to conduct interviews and to gather material. Now I realized two problems: one was that the pre-history of the ERC, that is, until the formal legal adoption of this instrument in late 2006, was much more intricate than I had expected. The other was that my first outlines for the book read like a clumsy advertisement for my study object, not like an illuminating analysis. I was too deeply entangled in the ERC's own perspective on things. To make sense of what had been going on and to distance myself would be crucial for the book's becoming an honest and serious attempt to narrate the ERC history. This distancing exercise went through several steps, not all of which were always purposefully achieved but rather happened to me (or so it seems in retrospect).

As contextual information it may be useful, therefore, for the reader to know how this book developed from a working relationship to a more scholarly interest. I had secured some funding for the time after the ERC, Helga Nowotny had helped me to broker a contract with a prominent British publisher, and I had gained a Fulbright scholarship for a five-month stay across the Atlantic to work at Harvard University. The result was a rough, 120-page manuscript about the pre-history of the ERC. This was a good exercise in sense-making, as it helped me to clarify where this creature had come from, what its ideational roots had been, who its advocates had been. Most importantly, it helped me to sort through the ERC anecdotes and get a good understanding of what happened when, where and why. I sent the manuscript to several of the people that I had interviewed before, and they replied with critical remarks but, in general, agreed that my account was fair and balanced.

Yet, after Harvard, and upon returning to Vienna, there was a serious hiatus in my writing. For one, I was taking over a new job that was also in science management but entirely different. But then, I didn't know how to move on with the book project either. I realized somehow that the ERC history would be incomplete if I were only to write on events pre-dating its formal inception; yet how to write its history since then was something that I could not frame. Maybe more to distract myself than intentionally, I started to look at the conceptual roots of exactly those aspects of the ERC that I had been concerned with when working there: the principle of distributing funds for research, the unsettled issue of autonomous governance, the quest to prove impact.

Again, my initial approach was naive: I thought that, if I could understand those concepts theoretically, I could also explain how great the ERC was. Only by reading studies investigating the historical traits of each of those issues did I realize that the ERC, as any other research-funding instrument, rests on a series of assumptions that have been developed over a long period of time and become locked in the collective consciousness of researchers today. Yet what seems to be embedded and fixed is actually continuously negotiated and fine-tuned. The most difficult step for me was to bring those concepts into a relation to the ERC operations, not only because I was working through the entire documentation of ERC Scientific Council sources, but also because I was still often not very sure where this journey would lead. Only very gradually did the book emerge as it stands now: a study of the conditions and constraints under which the distribution of public funding for academic research is to be organized successfully.

My publisher asked me, somewhat ambitiously, to write the complete history of the ERC. The expectation is understandable: I have been immersed in the ERC for almost four years, and I have spent another two years since primarily making sense of this instrument. Unfortunately, I still cannot claim to present the complete picture here: parts of the story that I examine remain incomplete, due to restricted access to sources – particularly, while I had by default the documents belonging to the realm of the Scientific Council at my disposal, I had very limited access to the Commission and the ERCEA side.[5] True, I talked at length with leading Commission representatives and, to a much lesser degree, also with people from national ministries involved in the early phase of setting up the ERC. I also had quite a lot of informal exchange with the colleagues at the ERC Executive Agency with whom I was involved during my time at the ERC. But where I can really claim full coverage of sources is only on the ERC's Scientific Council. So, while there will be things in this book that may be corrected in later academic work, I believe that this is the most comprehensive analysis on the ERC history that has been written so far.

Unlike other historical accounts focusing on a single research funding organization and its establishment, this book is not a 'court history'[6] (in the sense of an apologetic account) – not only because, in the case of the ERC, there is no court, but also because I attempt to look at the ERC from a critical perspective. To be critical does not mean that I have a revisionist agenda to debunk the alleged achievements of the ERC (or, more accurately, those speaking for the ERC). It does mean, however, that, unlike other insiders who have been publishing their account of the ERC history, I claim – both conceptually and methodologically – to be more rigorous than they in their rather anecdotal accounts.[7]

In order to do so, I treat my study-object consequently as something from the past, even though some of the evidence is still very recent. I am looking at the entire period, from the time when the ERC debate took off in the early 2000s to the time when the actual instrument concluded its first programme cycle in 2013. I try as much as I can to found every explanation of an event, and every relation between events, on written sources (that is, contemporary documents composed at the time of the event). And I make use of the plenitude of anecdotal stories that surround the various steps of the ERC's history (and its earliest phase in particular) only very carefully: I take them as evidence for the necessity to explain something, but I generally doubt their capacity to do so satisfactorily unless other (written) material proves them right.

The most fruitful and satisfying work in writing this book was the puzzle of putting together different events and arriving at a new understanding of their relation. Still, as much as I have tried to rest my account of the ERC history on rigorous assessment of written evidence, I cannot escape the potential criticism that this narrative, too, is an imaginary account of real events. I am aware of this inescapable trap; I cannot resolve it but can only ask the reader to judge, based on the evidence, the credibility of my remarks. To that end, I find solace in a quote by Hayden White: 'How else can any "past", which is by definition comprised of events, processes, structures and so forth that are considered to be no longer perceivable, be represented in either consciousness or discourse except in an "imaginary" way?'[8]

An important feature of the exercise to distance myself from the ERC thinking was also to avoid the typical Brussels lingo. Even where I could not entirely do away with abbreviations and shortcuts, I resorted to what I think is a more accurate form of speech. I speak of the 'Framework Programme format' (or 'FP format') when I refer to the EU 'Framework Programme for Research and Technological Development', in order to emphasize that it is a political instrument dedicated to covering EU spending for supporting scientific research and the development of new technologies. This format is to be distinguished from its actual multi-annual editions; what is commonly known as 'FP7', for example, I here call more accurately 'the seventh edition of the Framework Programme'.

A common figure of speech among policy-makers and scientists is to speak of 'frontier', or 'basic', or 'fundamental' research. While I respect the fact that those are broadly used terms, I have come to the conclusion that they serve primarily a political purpose, but carry little analytical value otherwise. The most accurate term for me seems to be academic research (vs. applied research), which simply refers to the fact that this is research conducted at academic institutions, such as universities and public research facilities.

The ERC, too, has created its own lingo: the 'Starting Grant' and 'Advanced Grant' are specific funding calls, the 'Scientific Council' is the independent body steering the ERC compound, the 'Dedicated Implementation Structure' (DIS) is the legal term for what would later become the 'ERC Executive Agency' (ERCEA), and so on. I have tried to explain each of them when introducing it for the first time, and also to use neutral expressions whenever possible.

I try to express as clearly and accurately as possible the names of European institutions and their organizational subdivisions, such as the European Commission, and its Directorate General (DG) for Research, or the European Court of Auditors; in the former case, I also use the shorter notion of 'the Commission' when it is clear what I mean by that; in the latter case I resort to the official abbreviation (ECA). 'Competitiveness Council' refers to the sub-group within the 'European Council', that is, the regular gathering of ministers responsible for trade, economy, industry, research and innovation, and space from all EU member states.

The political realm in which the ERC is located, and on which this book is focused, is innovation policy; in my understanding, it describes fairly accurately what political scientists call a 'macro-policy objective' of the EU.[9] Science policy, by which I mean exclusively, for the rest of this book, 'policy for science',[10] is part of innovation policy; and so is research and development (or R&D, in the language of insiders), which deals with the issue of how to stimulate private investments and distribute public funds to scientific research.

The field of innovation policy is not only full of abbreviations, but also of declarations, undisputed assumptions, metaphors. If I have made use of linguistic expressions typical of the social groups that this book is dealing with, I have done so only by putting them in quotation marks and, except in very few (and self-explanatory) cases, also referring to the source from where this is actually quoted. I have applied a similar approach to the chapter titles: to highlight that, ultimately, my narrative of the ERC story is as close as possible to the historical sources, they consist exclusively of quotes that have been used somewhere in the respective chapter.

— 1 —

THE FUTURE OF SCIENTIFIC RESEARCH IN EUROPE

On a sunny, mild Tuesday in October 2005, Slovenian politician Janez Potočnik addressed a small audience in the University Foundation of Brussels. Potočnik, then still a rather fresh member of the European Commission's political cabinet, the College of Commissioners, had taken up responsibility over the science and research portfolio six months ago. Now he was standing in front of twenty-two eminent scientists and scholars from across Europe, all of them highly decorated in their fields (including three Nobel Laureates), and many of them also long-standing advocates of science funding in Europe. Before that summer they had been invited to become members of the 'independent Scientific Council' that was about to steer an exclusive new funding instrument at European level, oddly called 'European Research Council', or ERC.[1]

The ERC, as the Commissioner briefly outlined it in the speech, would become a smallish part of one of many of the European Union's policy instruments, namely its research funding programme (in legal terms, this instrument is called the Framework Programme for Research and Technological Development). Of course, that's not how the Commissioner pitched it. One of the necessary skills of experienced politicians and their speechwriters is to find the right tone for each of the many occasions where they are asked to contribute some meaningful scores. In front of eminent researchers, Potočnik seemingly did not want to appear timid. Right at the beginning he called the meeting 'historic', and, a bit later: 'I can safely say that at stake is the future of scientific research in Europe.'[2] This kind of language pleased his handpicked audience, many of whom had been actively engaged in campaigning for the very organization that was now formally introduced to them.

Today, there is almost uniform consensus among social scientists and policy-makers on the assumption that 'economic growth is fueled by upstream research – research that is years away from leading to new products and processes'. Once this assumption is accepted, however, it poses a political problem, because new knowledge is a public, non-rival good (to speak in terms of economics) with the 'potential of having multiple uses'; consequently, there are no 'economic incentives' for 'any one individual, company, or industry' to support it.[3] On the whole, innovation policy aims to solve this perceived problem and to achieve economic growth (as well as other socio-economic benefits) mostly by distributing public funds to research and by making sure that the new knowledge produced through this research is transferred to the marketplace.

Historically, the importance of scientific research for economic growth can be traced back to two developments in the second half of the twentieth century, one being ideational, the other institutional.[4] The first is an opaque concept around the term innovation becoming a major study object on its own in economics as well as in other social scientific disciplines.[5] At the same time while its meaning was considerably narrowed down to, more or less, technological advancement, the public attention to it made it near to 'the a priori solution to every problem of society'.[6] Of course, those two trends were deeply intertwined, for it would not have occurred without the work of social scientists, and economists in particular, that the idea of economic growth based on knowledge (transferred in technology) would have become a political mantra; and, without the links to public policy, 'innovation studies' would not have become its own academic tribe.

The second development was closely related to the new belief: ever since the United States discussed, and eventually created, the National Science Foundation (NSF) with the intention of enabling scientific research at universities, the link to innovation was only increasing. In one of the Annexes to the constitutive report 'Science – the endless frontier', it predicted that '[i]n the next generation, technological advance and basic scientific discovery will be inseparable; a nation which borrows its basic knowledge will be hopelessly handicapped in the race for innovation'.[7] Consequently, the distribution of public funds became a key factor in the relation between science and policy (albeit not the only one).

There exist different convictions of how best to use public resources in order to stimulate innovation, and what kind of investments would yield the best results; those convictions, despite their differences, usually have two common strands. One is that, while they are often based on well-reasoned arguments, and sometimes even analytical models and economic theories, once they are brought into the arena of innovation policy, they can hardly be separated from the vested interests of those speaking for them. The second is that they rely on a catalogue of labels such as 'basic', 'applied', or 'frontier' research, or 'interdisciplinarity', or even 'innovation' itself.[8] Those are labels that, by default, cannot fully do justice to the complex and highly diverse set of practices carried out in very different situations (academic, profit-oriented), spaces (labs, libraries, research facilities), and along different disciplinary as well as epistemological precedents.

Over the decades, however, the priorities of the policy discourse have been changing quite substantially.[9] If, in the beginning, it was generally believed that funding 'basic' (i.e. academic) research would do the trick, governments would later concentrate on more strategic investments; and from the 1990s onwards, it was common to speak of a portfolio of diversified instruments to foster innovation from research carried out at universities to give support to industries and businesses.[10] An important consequence is that innovation policy is diverted into different subsets of secondary goals supposedly tackling different aspects that contribute to (that is, help stimulate) innovation.

While common-sense today, this upscale version of the shotgun-approach indicates a specific problem of innovation policy (and, more specifically, the distribution of funds for scientific research): effective input. Innovation policy is not alone with problems such as that there are unintended consequences (that is, regulatory or distributive policies will be made use of in ways that cannot be foreseen), or that public funds are a scarce resource, as demand is always higher than supply (science is expensive, but so is defence, social welfare, and so on). However, what is so difficult to pin down is how to actually make innovation happen; or, to be more precise, how to make people (or companies, or businesses) innovative. In other words, the realm of innovation policy consists of a plethora of programmes, instruments, and institutions, most of which distribute public funds in a way that is based on a certain mission and the service of their particular clientele.

In the case of European integration, research funding has been an undercurrent for quite a while,[11] and in the broader contest of the so-called Lisbon Strategy in the year 2000 it gained special attention. The Union probably never came close to achieving the long-term goals set out in that document; nonetheless, the strategy had been 'an extraordinary process of intellectual mobilization across Europe and beyond'[12] and remained of high political significance. It was in the light of this strategy that the vague idea of an independent funding body named ERC was becoming a tangible goal for a group of fervent, influential advocates; and 'the well-known objectives' of the strategy were a reference point that Commissioner Potočnik could not miss in his speech to some of those advocates now[13].

Research funding takes a special role in the entire portfolio of tasks that the European Commission is required to fulfil. One reason is that its volume is fast growing, with the latest edition of the Framework Programme for Research and Technological Development (FP), called 'Horizon 2020', having € 80 billion committed over seven years (2014–2020). Around the time of Potočnik's speech, political negotiations for its predecessor, the seventh edition of the Framework Programme (commonly known as 'FP7'), were in full swing and the European Commission could make good use of the ERC to argue why it asked EU member states to increase its overall budget. In the end, the seventh edition was at approximately € 54 billion from 2007 to 2013; between 15 and 17 per cent of that budget was allotted to the ERC, and all of it was 'fresh funding', as the Commission had demanded.[14]

The numbers are dwarfish in relation to other significant EU policy areas of (re-)distribution, such as the agricultural subsidies (Common Agricultural Policy, or CAP) or the structural funds. Other than them, however, the research funding budget is managed directly by the Commission by 'allocating grants to private and public beneficiaries'. This requires heavy machinery and comes with risks of being politically exposed for wasteful spending. As a report on the European Commission's annual fiscal conduct during the phase of establishing the ERC, ingenuously put it:

> The principal inherent risk to the legality and regularity of the underlying transactions [of funds, TK] is that beneficiaries may overstate costs in their declarations, partly due to complex legal and contractual provisions, such as insufficiently clear definitions of eligible costs. A lack of sufficient sanction mechanisms amplifies the risk.[15]

In this context, the European Commission also hoped to use institutional innovations like the ERC in order to streamline its administration and to outsource a good part of the funds it was supposed to distribute.

For an international organization like the European Commission that envisions itself as a policy-making institution, it is much more attractive to oversee regulatory measures than to disperse billions of euros and be held accountable for it. The rise of innovation policy around the millennium indicates the Commission's appetite to actually extend its policy portfolio. The most articulate expression of this ambition was the 'European Research Area' (ERA). This was an attempt to build up a more comprehensive framework to overcome 'fragmentation, isolation and compartmentalization of national research efforts' and to achieve 'better integration of Europe's scientific and technological area'. A research funding instrument specifically dedicated to fund academic research would, potentially, also offer 'a European dimension into scientific careers', provide a lever to establish common standards, and make 'Europe attractive to researchers from the rest of the world'.[16]

Increase of budget, improving its funding-regime, and extending its policy portfolio were not the only reasons why the European Commission embraced the ERC, and Potočnik was not completely wrong with his aggrandizing rhetoric. Ultimately, the ERC was also to become something of a flagship for the next edition of the Framework Programme, which had often been criticized by scientists for its funding programmes being 'Loch Ness monsters of bureaucracy'.[17] Against this backdrop, the ERC was supposed to '[g]et the aura back', as one of the politicians involved in its creation openly stated.[18] It was this promised aura that Potočnik expected to be fully supported by the advocates now assembled in the room in front of him. After all, they had rallied behind the ERC campaign exactly for that reason.

When they were convened in that room in October 2005, advocates and Commission representatives not only shared a common belief in the future ERC's aura, but also in the two basic principles that would make sure that this aura could be established and maintained. The first principle was that the ERC would be commissioned to fund research projects regardless of nationality; in ERC-speak, this approach would soon be dubbed 'excellence only', indicating that no other criterion than the scientific quality of a project proposal was to be taken into consideration for decision-making. This focus on 'excellence' certainly was the unique feature of the ERC, at least in the context of research funding at European level. ERC funding calls became very attractive to scientists, not only because of the size and the competitive nature of the grants, but because of the reputation that was going along with them.

The other shared principle was that the ERC would be headed by the Scientific Council, the group of people in front of the

Commissioner. This body would have the right to establish 'the overall strategy' for the instrument, in particular to prepare 'the work programme', to come up with 'the methods and procedures for peer review and proposal evaluation on the basis of which the proposals to be funded will be determined', and to establish 'its position on any matter that from a scientific perspective may enhance the achievements and impact' of the ERC.[19] No other council, committee or advisory board in the realm of the European Commission had ever been granted such extensive rights.

It should be added immediately that the two principles on which the ERC was expected to found its aura were nothing new to the world of research policy per se, since many industrialized nation-states have similar funding instruments resting on exactly the same two principles; they would be unique only in the context of the supranational context of European integration. However, the long absence of those principles from EU research policy had its well-founded reasons, and it is all the more interesting to ask why, over a relatively short instance, those reasons could be overruled. Furthermore, if the ERC were a serious attempt to emulate these long-standing principles at transnational level, it would have to come up with a few peculiarities on its own.

Were those gathered in the room with the Commissioner correct? Could the ERC, once established, keep its promise? The wide-spread (if not unanimous) opinion held by policy-makers and scientists across Europe is that the ERC is a success, or, even more pronounced, that it is a 'European success story of European integration'.[20] Even José Manuel Barroso, not necessarily its most fervent advocate during his reign as President of the European Commission, counted the creation of the ERC 'among his greatest hits' at the end of his final term.[21] This notoriety is quite remarkable for a tiny funding instrument. The common belief and constant rhetoric of a 'success story' indicates clearly that the expectation of the ERC bringing 'added value' not only to the scientific research in Europe,[22] but also (more selfishly) to European research policy (and its leading personnel), has been achieved.

Having been involved with the ERC myself for quite a while, I know about the power of its ideational fixture, not only in tying together different legal entities but also in providing the ERC with a symbolic surplus that goes beyond the resourcefulness of this policy instrument. This book, then, is about the ERC's aura. However, success stories can boldly be claimed in speeches and public presentations; but, while readers may come to the conclusion that the ERC story is one of success (or one of failure), and while I will happily return to discuss this in the final pages, for the remainder of this book the issue deserves a subtler handling.

There have been a few attempts to analyse the ERC and its creation in particular, either as an example of institutional evolution and improvement in the EU, or as a subject of discourse analysis, or as a case study of principal-agent-theory.[23] Other than those investigations, the present book follows a more holistic perspective on the ERC, taking it as a case study within the unique transnational realm of European integration to investigate three key issues that can be brought down to the following questions: how was the ERC's aura cast? How was it institutionalized? And how was it routinized? It is easy to grasp that, while those questions result from the ERC's own history, they also relate to a broader framework, namely the intricate relationship between science and policy.

While this book does not strictly follow a specific theoretical thread, it still claims to be more than a reproduction of events. To do so, it draws on literature in the tradition of a pluralistic approach in EU studies,[24] on the one hand, and of science and technology studies, on the other hand.[25] Despite conceptual differences within each of these two strands, they share the same underlying epistemological rationale, which I would call 'constructivist'. My main interest, thus, is not whether the ERC had been a long-absent necessity in the European research funding landscape, but rather how its advocates had been arguing for this instrument and how it was brought through the political procedures at European level (Chapters 2 and 3). It is not whether it was a missed opportunity in respect to its institutional arrangement, but rather how this arrangement had led to different expectations, tensions, and compromise (Chapters 4 and 5). And it is not whether the ERC is funding the right research, but rather how its distribution of public funds is legitimized and reasoned for (Chapters 6 and 7). In other words, mine is an attempt to take a deeper look at what is actually going on inside an institution deliberately erected at the borders of science and politics, and to grasp the messy situation underneath the glossy world of research funding policy.

The story of how the ERC came about (or, what I also refer to, the ERC pre-history) concerns how the promise of aura became so politically convincing that the wild dream of an independent European Research Council was realized, against all odds, in legal texts and budgetary commitments. This story starts with the campaign of a group of scientists and science managers pushing for the creation of the ERC; Chapter 2 sets out to identify how the ERC idea, an old

	Traditional rationale	**Complementary (ERC) rationale**
Targeting	'Pre-competitive' research	'Basic' (later: 'frontier') research
Achieved through	Cooperation, mobility	Competition
Added value	Transnational collaboration	Scientific excellence

Figure 1.1 Traditional and complementary rationale of European research funding

Compiled by the author.

but unrealistic dream among senior representatives of the scientific communities in Europe, turned into a powerful yet loosely connected campaign. For a while, then, this story is one of a grass-roots movement, whose members were grappling with, on the one hand, aligning the scientific communities across Europe, and, on the other hand, making themselves heard by politicians.

The story takes an important change in tone when, a couple of years later, the ERC idea was lifted into the political arena of the EU. Chapter 3 shows that, when the European Commission decided to subscribe to the ERC, it altered the campaign into a much more strategic, planned and technical approach, which included the invention of a new, complementary rationale on justifying funding for research at European level (see Figure 1.1). This was certainly necessary in order to carry the proposal through the complex political process of the European polity. Yet the alliance between the ERC advocates and the Commission leadership that was created upon this move was not without its frictions.

Chapters 4 and 5 deal with how those in charge of the newly incepted funding instrument grappled with realizing a governance structure that was adequate for the needs of implementing and maintaining the ERC's promised aura and yet in accord with the requirements and provisions of the European polity. This part of the book, then, looks at how the ERC was actually implemented. Eventually, political compromise between ERC advocates and the European Commission leadership resulted in the creation of the ERC's Scientific Council, which was assigned an extraordinary realm of autonomy and which was put together of individuals who were not only highly distinguished scientists and scholars, but also fervent advocates of the ERC as a separate instrument to advance academic research across Europe. Remarkably, the group was created more than a year before the formal inception of the ERC under FP7, thus enabling it to work into the void and set up the ideational framework as well as the operational details of a seemingly independent, science-oriented funding machine. The first fifteen months of the Scientific Council's existence were indeed a formative period for the ERC, and a particular one of the ERC's pre-history, which is framed in Chapter 4.

The ERC had been dreamt up as a European copy of the NSF, and taken only from indicative numbers, it may have appeared as one; the reality was different, however (Figure 1.2). Since the EU polity is organized in a completely different way from that of the United States of America, it also provided different institutional templates, and came up with a strikingly different rationale for the ERC's guiding governance principle. Most notably, despite the Scientific Council's impressive appearance, the Commission would also carefully circumvent the position of this body. As a consequence, the ERC compound was torn between two ideas of how to run a funding instrument. The frictions around governance of an allegedly independent structure within the European Commission are being looked at in Chapter 5.

As a funding instrument, the ERC was expected to blast huge amounts of money into highly specialized research projects. This required heavy and specialized machinery, one part of which would be directed at making the funding allocation, and the other at establishing contractual conditions for the actual payments. The last part of this book, investigates how the promised aura was established,

Items	NSF[26]	ERC[27]
Budget (approx., in bio.) / Staff FTE	$ 7.1 / 1,400	€ 1.7 / 380
No. of proposals submitted/ funded	50,000/11,000	10,000/1,000
Legal status	Independent Agency	Compound of 3 legal entities
Funding	US Congress	FP format
Regular funding streams per annum	Up to 300	4
Decision-making principle	Peer review	Peer review

Figure 1.2 Comparison of NSF and ERC on various items, 2014

Compiled by the author; see endnotes for references.

solidified, and routinized in the ERC's core decision-making procedure and in its increasing relevance to research institutions and researchers across Europe.

In total, the ERC would run eighteen major funding calls between 2007 and 2014,[28] along which more than 5,300 research projects were funded, based on the assessment of about 50,000 submitted proposals (Figure 1.3). It took a while for the machinery to be in full operation, but by the end of 2013, the ERCEA employed almost 400 staff and the ERC's selection procedure required the input of approximately 5,600 reviewers.[29] Behind those numbers stood (literally) tens of thousands of ideas for research projects that had been submitted and needed to be sorted, evaluated, ranked and decided upon; and the projects that were actually funded had to be negotiated, contracted, supervised, sometimes extended, sometimes audited and, at the end of their duration, terminated. This required a plenitude of tasks, decisions and consecutive steps, all of which had to be defined, dovetailed, monitored and carried out with due diligence, based on a set of certain rationales and guidelines, converging scientific, political and administrative logics. What tied them together at the most fundamental level was the selection

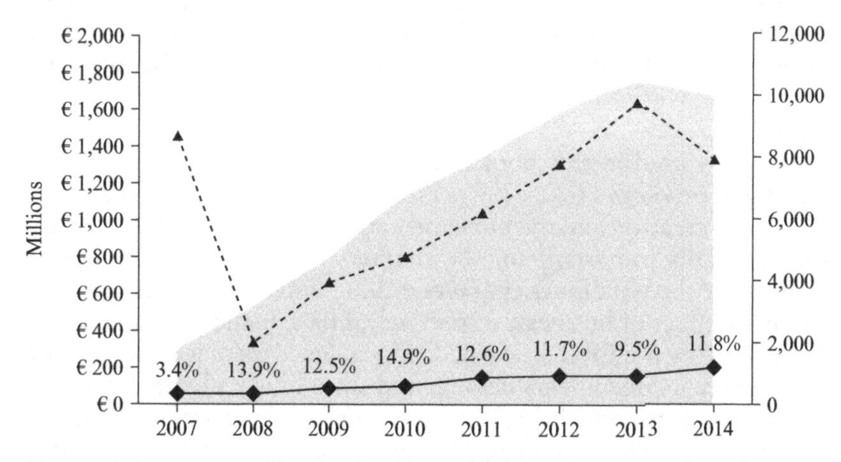

Figure 1.3 ERC indicative budget, proposals submitted, and grants funded, 2007–2014

Compiled by the author; data from ERC Work Programmes and Annual Activity Reports. Annual numbers (not including top-up and support grants); grey area: annual budget dedicated to research funding; dashed line: proposals submitted; continuous line: grants allocated; numbers: overall annual success rate of submitted proposals.

principle: to allocate funding based on the scientific quality of a proposal, judged by independent, competent, and impartial reviewers.

The most striking aspect of the ERC is the fact that it was able to brand a decade-old feature of the scientific community as something innovative and new. The funding philosophy was based on the principle that only intrinsically scientific qualities should be decisive for the allocation of funds. For centuries, peer review (the name of the principle) has been an integral part of the 'scientific culture';[30] its application to the allocation of public funds reaches decades back to the post-World War II period. As Chapter 6 shows, the way this principle was implemented, and the way this implementation was orchestrated, however, would require exact fine-tuning and adjusting different expectations into the ERC – political as well as procedural ones.

Chapter 7 returns to the issue of innovation policy at EU level and investigates the ERC's continuing quest to prove its value and to remain a positive fixture in the political debates a few years after its inception. While the ERC was widely praised for being so successful, proving this success would turn out to be not so easy, after all. Particularly how to show the impact of this funding instrument would claim the attention of the ERC leadership. Yet despite these problems, the chapter also provides some evidence that the relevance of the ERC within the European landscape of research funding was significant.

The data used in this book are taken from different sources. For one, I was privileged to use the personal archives of several key actors during the creation and the implementation of the ERC. These sources were crucially important for the chapters on the pre-history in particular. For the remaining chapters dealing with the ERC governance and its conduct of business, respectively, I have made extensive use of the archive by the ERC Scientific Council. Since there is no regulation as to where its documents belong, and since I have obtained them as part of my job with the Scientific Council, I deliberately made use not only of public documents, but also background papers, internal presentations, statistics and confidential material. The only material I left out (except in a few cases where I explicitly asked for permission) were emails.

Furthermore, I was conducting interviews and did a good deal of research on public databases (most notably, the European Commission's publication and press release database). Another important source, particularly for circumstantial evidence, was – what could be called somewhat disrespectfully – the yellow press of international science policy affairs, namely the comment pages of (otherwise highly esteemed) journals such as *Science Magazine*, *Nature*, *Cell*, and so on. Those pages are traditionally read by scientists as well as by science policy-makers in Brussels, which is why their (extensive) reporting on the ERC offered a great source of material for contextualization.

There are two types of source that I did not make use of systematically. For one, while I was at the ERC I was using a notebook – and had I decided earlier to write this book, it would probably contain a set of invaluable field notes. Instead, my remarks for much of the time are mainly summaries of talks and meetings, and agendas for structuring my daily work (only lately did I start to write down observations and reflections upon what I had witnessed). Another source would have been a survey with all chairs of ERC evaluation panels between 2007 and 2013. The survey was thought to provide another systematic avenue to the analysis of the ERC's peer review procedure, and I am extremely grateful that more than 50 per cent of the 158 chairs who are extremely busy scientists and scholars replied to my questions. However, while I found much detail in the replies, helping me to fine-tune my analysis, I ended up by not exploiting the data in a systematic manner.

This book is primarily interested in politics – or, to be more precise, the way the ERC was conceived and implemented, and the way its aura was, first, fashioned to a political argument, and then established, reinforced, routinized. It will tell almost nothing about the specific research that has come out of funding provided by the ERC. And, since it is mostly concerned with the ERC, it will not analyse systematically the state of current innovation policy (or any of its branches) in Europe, or the relationship between science and policy in general. Yet I believe telling the ERC story from a critical perspective is worth telling for three reasons, and that it should be interesting for various kinds of people.

Research funding instruments, such as the ERC, could be perceived as an individual species within a distinct class of giant insects: dispersed across the globe, they still have striking commonalities. They share the same purpose, namely, to distribute large amounts of public funds to research; their activities have to be credible to scientists as much as to policy-makers;[31] and they all follow the same functional morphology: their organizational skeleton consists of three distinctive but interconnected units, with the head for strategy and monitoring, the thorax for operating the extremities and making funding decisions, and the abdomen for contracting, auditing, and all other activities concerning the organization's life support.[32] Despite those commonalities, physical appearance of funding instruments is greatly differing, since each of them occupies an ecological niche in its respective 'innovation system'.[33]

Against this backdrop, the creation of the ERC is quite remarkable, with a few nice twists and unexpected changes, and it would be a pity if it were to be forgotten. Veteran ERC advocates and anyone interested in the creation of the ERC and how it came about will find plenty of material in the following pages. Writing a comprehensive story always includes the need to omit certain details and, inevitably, it may even reveal a bias in the judgement of events. It would be a great reward if this book would contribute to a discussion about the historical traits and the political developments in the ERC pre-history, and about the ERC's achievements in establishment and maintaining its unique aura.

Second, the ERC story is also a tiny yet significant part of a certain period in the 'continued political experiment' of European integration,[34] and of the intricate relationship of science and policy during that time. To recollect what was then deemed politically important and relevant is not just to reflect on the general mood of the period, which appears (from today's perspective, at least), to have been optimistic, yet also a bit naive. The campaign leading up to the creation of the ERC and the implementation of the policy instrument offers plenty of insights into the depths of European politics and bureaucracy in the field of science policy, and this book contributes to the on-going scholarly discussion around those topics.

A third reason for telling the ERC story is that it also reveals a lot about the relationship between science and policy in the twenty-first century more generally. It is about successfully unlocking additional public funds for academic research, but also about the conditions under which such a policy instrument is to be set up today – the challenges to the relative autonomy of its strategy-devising leadership, the need to orchestrate its decision-making procedure, the need to prove impact. To that end, this book addresses scholars in the field of innovation policy and science and technology studies, but, more broadly, it also speaks to research managers and policy-makers. In much of the literature on either of the latter three topics in particular, there is too little reflection on the theoretical basis and historical traits of those concepts, and hence they are usually treated too much as unquestionable givens by scientists, scholars, research managers and policy-makers alike. Ultimately, thus, this book would like to contribute to a process of scrutinizing these givens, in order to avoid that they turn into – what social theorists call – an ideology.

— 2 —

A RADICAL PROPOSAL

The idea of a 'European Research Council' goes way back to the 1950s,[1] but it was only from the year 2000 onwards that claims for establishing the ERC received serious attention for the first time. Even then, nobody would have foreseen how quickly the ERC would come to reality, nor that it would be established under the command of the European Commission. To understand the nuances of the ERC debate starting in earnest, and the broader context of the ERC campaign unfolding from there, it is necessary first to have a look at how investments in (what was then called) 'research and technology development' were discussed and managed throughout the 1990s. Mainly, this was about the so-called 'European paradox' and the frictions around the central policy device, the Framework Programme for Research and Technological Development (FP).

With the Lisbon Strategy and the concept of the European Research Area, the debate intensified; at the same time, a counter-narrative emerged that was directed against the principles and, even more importantly, against the FP format. The counter-narrative quickly became the first building block of the ERC campaign that eventually gained a foothold among ambitious representatives of the scientific communities across Europe. This chapter discusses how the ERC idea, then merely a construct of different shades and shapes in the minds of different people, would gain enough momentum to be ready to jump into the political realm. The following, thus, will also clarify some of the underlying tensions that would continue to trouble the ERC later, as well as the ideological foundations of that instrument that will help us to understand its key features.

2.1 'Not-yet-born' sector actors

When the Lisbon Strategy was adopted in 2000, it pushed for Europe 'to become the most competitive and dynamic knowledge-based economy in the world'.[2] This was an important political milestone for the European polity as a whole; it was also a consequence of the political and intellectual zeitgeist of that time.[3] Since the 1980s, research and development (R&D) had grown to be an important policy feature of European integration. In part, this was due to a loose set of new intellectual concepts emerging during that period that also fed directly into the Lisbon Strategy.[4] Stimulated by concepts like 'globalization', 'innovation', and 'new production of knowledge', scholars began to think beyond the sovereign nation-state; they enlarged their understanding of innovation from a narrow *terminus oeconomicus* into a core mechanism of capitalist society; and arrived at a new perspective on the relationship between science and society.[5]

The optimistic spirit was supported by the seemingly unstoppable political progress: witnessing the fall of the Communist regimes in Eastern Europe and the accession of five additional countries (Spain and Portugal 1986, and Austria, Finland and Sweden 1995) with a dozen others to join soon, the European Community changed not only its name (since the Maastricht Treaty in 1992, the polity called itself a Union), but also its complexion. Becoming more and more synonymous with Europe,[6] it had considerably enforced its integration measures through several landmark treaties, from the Single European Act in 1986 to the Amsterdam Treaty in 1999. In all of them, research policy gained more and more prominence. The political consensus that R&D would be vital for the European integration process and for the future wellbeing of its citizens, and societies as a whole, brought in a new momentum. Scholars and policy-makers alike started programmatically talking about the 'European science space', and publications on its history and present formation appeared.[7]

One focus of the growing funds for socio-economic research was on foresight studies, and 'science and technology policy options' in particular.[8] Identifying what was called, in the inimitable European Commission jargon, '"not-yet-born" sector actors'[9] was probably as ambitious as it was futile, but it erected a place where policy-makers and scholars (mostly economists and sociologists) met and discussed the imperatives and impediments of European research policy. The papers and concepts, conferences and workshops from those projects fed into the growing discussions about how to best spend public funds for research in Europe.

Thanks to the mixture of historical explanation and the powerful imagination of a geopolitical space, available statistical evidence and demand for political advice, the specific issue of R&D developed its own narrative. Despite the politically upbeat atmosphere, the mood expressed in this discourse was distinctively darker. One reason was the rather disturbing figures showing an ageing population and a growing unemployment rate, and economic shifts within the Union as well as a general decline in growth. But what really caused the R&D scholars and policy-makers a headache was their perception of a polity being unfit to react appropriately to those demands. Statistics produced by OECD and Commission services indicated that the 'emerging post-national system of innovation' in Europe[10] was lagging behind the two other economic world powers, the USA and Japan. Indicators were the 'relatively small number of researchers', the lower 'number of university graduates per year', lack of 'public investment in education and human capital', and also the 'lower proportion of its resources to S&T', and its overall 'technological performance'.[11] The widely read Commission Communication on the ERA stated in its introductory note that

> the situation concerning research is worrying. Without concerted action to rectify this the current trend could lead to a loss of growth and competitiveness in an increasingly global economy. The leeway to be made up on the other technological powers in the world will grow still further. And Europe might not successfully achieve the transition to a knowledge-based economy.[12]

The problem, as the architects and scholars of R&D saw it, was how to devise a policy at European level that would live up to expectations – how to coordinate and support R&D. Quite naturally, then, its major policy instrument took centre stage in the discourse.

Research and development had been present at the level of the European Community since its very inception in different sectors and as different institutional manifestations;[13] attempts to evolve 'from sectoral to "general" research policies'[14] can be found as early as in the 1970s.[15] While those attempts to gain more regulatory power at the European level were effectively rejected by the member states, the distributive means of different 'research, development and demonstration (RD&D) activities' were finally incorporated under the first framework programme in 1984. According to a Commission communication of that time, the unified framework would promise that 'resources will no longer be handed out piecemeal according to the circumstances of the moment; their distribution will be governed by a political determination to do everything possible [. . .] to meet the challenges of the future'.[16]

The first FP entered the stage at a time where the European research landscape was already quite crowded with other instruments such as COST and a number of European research institutions such as CERN or EMBL.[17] In the beginning, it was probably not much more than just the rebranding of 'all the separate research and development programmes' within the Commission services.[18] But the FP format was different to the aforementioned tools and institutions, as it put the distribution of means into the hands of the Commission, and it retained the vague hope of exerting some indirect power to coordinate national research policies through funding leverage.[19] In these ways, the FP format changed the terms of reference for R&D policy across Europe, and according to some, it brought in the 'institutionalization of EU research policy'.[20]

For a while, the instrument proved to be well suited to the complex process of policy-making in the European Community. It was flexible both towards thematic priorities,[21] and in the scope of research that it funded – the general term employed here was 'pre-competitive' research,[22] a term that 'rules out very little'.[23] Because it spanned several years, there was enough room to react towards new political requirements while maintaining some sort of stability of priorities for the main recipients of its funds. The FP format, in other words, had its merits, and as such, it gained in political importance and became embedded in several of the European treaties. The Maastricht Treaty even specified that 'all the [research, TK] activities of the Community' would have to be bundled within the FP format.[24]

Yet, as scholars and policy-makers observed with growing frustration, it could not keep pace with the developments in Europe and the world. In the 1990s, the FP format had gone through several editions and had created its own effects.[25] One effect was that, instead of devising the instrument after the problem, the instrument increasingly dictated how scholars and policy-makers constructed the problem. It reinforced the 'so-called European paradox'; through statistical evidence, reports and articles, the notion held 'that the EU's healthy performance in scientific research is not being translated into strong technological and economic performance'.[26]

The second effect was that the FP format created serious institutional inertia: the distribution of funds tied up valuable administrative resources of the Commission, the complex regulations of the Union blocked attempts to change its procedures, and 'self-reinforcing policy networks' resisted any attempt to change the status quo.[27] All through the decade it proved impossible to escape what the instrument defined as feasible and possible in regards to R&D policy.

Finally, the most critical effect was exactly that status quo: the instrument had become synonymous with European R&D policy in general. As the conventional wisdom had it, European research policy was a distributive measure with limited scope; it had to strengthen the Union's 'competitiveness, both agricultural and industrial/commercial'[28] through supporting and supplementing industrial research in order to achieve a socio-economic impact; it had to provide European added value through funding collaborative (i.e. transnational) research projects and, a bit later, also mobility within Europe. It was also expected to return to each member state approximately the same amount of money that it had put into the R&D pot (the informal principle usually referred to as *juste retour*).

The political success of the FP format can be measured in the growing number of objectives that were ascribed to each of its editions,[29] but also in its funding volume. By 1998, when the fifth edition of the Framework Programme was launched, it had an impressive total of € 15 billion (whereas the first FP had started at a modest 3.7 billion ECU); R&D funding now roughly accounted for more than three per cent of the entire EU budget (it had started with only a bit more than one per cent). But it had also grown into a complex structure of thematic and horizontal programmes, 'functions', 'types of activities', and 'key actions',[30] all of which had to go through political confirmation in the complex web of EU comitology. No wonder that, when it came to the question of whether the FP format could also fulfil the many expectations, the outlook was bleak.[31] As early as 1992, a report by the European Commission stressed that 'improvements in this area are necessary, in terms of both methodology and organization', and concluded with a 'reorientation of Community R&TD policy'.[32] Notwithstanding such efforts, the self-criticism was only reinforced by an influential report several years later, which started almost sixty pages of assessment with the admonition that 'the Framework Programme is not fulfilling its promise. It lacks focus and is underachieving.'[33] Three years later, a similar exercise stated 'that existing policy tools need to be further exploited in a restructured and expanded Framework Programme.'[34]

Such critical reviews served a specific purpose, of course, as did many other interventions on R&D matters. Policy-makers in the Commission and many innovation scholars were desperately looking for ways to overhaul the instrument. That, however, required walking a tightrope, by justifying the FP format in principle on the one hand, and, on the other, identifying the setscrews whose subtle manipulation would turn the instrument around. That is why today, much of the textual residuals of that time make such awkward reading: they were written in a highly technical jargon and remained politically inconclusive. At the same time, each new edition of the Framework Programme claimed to represent a radical break with the past; each was accompanied by a host of reports applying concepts from science and technology studies as well as from economics to prove that this funding was useful and addressed a need for the future.[35] The proclamation of the European Research Area (ERA) in January 2000 was only the latest of those attempts, trying to put the upcoming negotiations for the sixth edition into a broader political context.[36]

In the late 1990s, not only was the FP format changing into a complex web of funding programmes, policy ambitions, and vested interests, but the discourse around it started to diversify, too. All through the decade, new actors had entered the stage, motivated as much by the increase of European funds for R&D as by the opportunities that the bolstered idea of a unified European science space offered. Scientists striving for a democratic vision of European science; managers concerned about the relation between their national organizations and European R&D; social scientists fearing for their funding share; Eastern Europeans looking for collaboration; representatives from ivory towers reaching out for a new mission – some of it was idealistic, some opportunistic. Mostly, this had been a subtle process, with occasional flares of attention, such as the inception of EuroScience in 1997, a grass-roots organization, which combined many of those strands,[37] or setting up the European Life Science Forum (ELSF) in 1999.[38] Contributions of the new actors revolved around the FP format, too; but since they were not forced to make sense of it, they focused rather on identifying a politically viable alternative representing their wishes. An independent research-funding agency at European level had been a popular candidate for such an alternative all through the 1990s.[39] Loosely speaking, a counter-narrative to the mainstream discourse was evolving, with an ERC at its centre. But this counter-narrative was missing an alternative historical explanation mixed with statistical evidence that could turn the paradigm of the 'European paradox' around. It required a veteran of innovation policy research and European science policy discourse to accomplish this.

2.2 Maybe it is time for a European Research Council in some form?

Keith Pavitt was a Professor at the Science Policy Research Unit (SPRU) at Sussex University;[40] SPRU had long been one of the most influential think-tanks within European innovation policy. Like many of his peers, Pavitt had grown frustrated by the several editions of the Framework Programme; his suggestions were firmly within the margins of the economic mainstream. He believed that 'indirect policies' (i.e. regulatory power) would yield better results, as they 'can have a major influence on corporate decisions about technical change'; as indicated, the Commission aimed for more regulatory power in R&D matters, too. Pavitt was in favour of 'supporting more speculative basic research programmes and networks, especially at the boundaries among disciplines and institutions (including business firms)'.[41]

In June 2000, when Pavitt was invited to a workshop in Lisbon of a FP-funded project on 'scenarios for the evaluation of the European science and technology policy', the political context had shifted. Earlier that year, Commissioner Philippe Busquin (Potočnik's predecessor) had announced the European Research Area (ERA), and in March, the Lisbon Strategy was adopted.[42] Now more than ever seemed to be the moment to rethink the foundations of the entire R&D policy. Pavitt, thus, did not linger on the topic of that meeting; instead, he made a 'radical proposal' to establish 'a European agency for funding academic research'.[43] Not only was it held at a momentous time and place; it spoke in the technical language of innovation policy-makers; and it proposed the new body as an alternative to the instrument that was currently in place.

Pavitt's argument followed three steps. First, he demonstrated that the European discourse had it all wrong when it disdained academic research and its contribution 'to economic and social progress', and emphasized the 'central role [. . .] in the systems of innovation in Europe and the USA'. Next, he debunked the 'European paradox': neither was Europe doing better in output of its academic research, nor were R&D expenditures by European businesses declining across the board – instead, Pavitt suggested, 'European firms are performing an increasing share of their R&D outside their home country, and more specifically in the USA'. The reason for that was

> that the strength of US academic research is one of the factors causing European firms to increase the share of their research performed in the

> USA, particularly in pharmaceuticals and related biotechnology, and also in ICT.[44]

Finally, Pavitt concluded that the superiority of the US was due to the uncompromising funding of basic research through public funds – almost delightedly he destroyed the core basics of the FP format: instead of foresight, 'in matching long-term technological opportunities with economic and social needs', the US was successful because of the 'unintended consequences' of massive spending in basic research; instead of demonstrating 'practical usefulness and user involvement at the project level', 'usefulness was defined in broad terms [. . .] which allowed the development of both long-term, speculative and fundamental research programmes, and post-graduate training'; and instead of a '"democratic" spread of funding to many regions', 'funding tended to be concentrated in relatively few elite institutions'.[45]

None of the components of Pavitt's argument was new. His genius was to combine an old dream of European scientists (an independent European research-funding agency) with a fairly recent argument from across the Atlantic (acknowledging the substantial role ascribed to universities and public funding for basic research within the 'innovation system'),[46] to wrap it into the language of the European mainstream R&D discourse, and to release it at a critical point in time. Thus it became a powerful instrument in the hands of those critical of the present European innovation policy. Not only was its priority wrong, it was even potentially dangerous, as it focused on support for corporate R&D, while the same corporations transferred their own research departments to the US. If Europe were to become the most dynamic knowledge-based economy in the world, as the Lisbon Strategy had promised, it had to revise completely its way of approaching things and to bolster the fabric of its academic research.

The counter-narrative was born, but the message was not yet heard by policy-makers. Between Pavitt's article and the formal step to put the ERC on the Union's political agenda lay more than two years. How, then, was the momentum created by the article pertained and prolonged? And how did it actually make its way onto the political agenda? One important factor was passing on by word of mouth. Pavitt's article spread within the academic communities, along with its basic ideas. The plenitude of European events on R&D policy in the aftermath of the Lisbon Strategy provided a fertile ground for popularization; the topic of the meeting didn't really matter. Only a few weeks before Pavitt's presentation, UNESCO held another conference on 'European

S&T [Science and Technology, TK] policy and the EU enlargement', with members from Europolis, an Eu-funded foresight project. One of them, Pierre Papon, suggested in his talk a 'complementary scenario' in which 'funding access to European research infrastructures [. . .] might be handed to a "European Research Council"'.[47] Papon then most certainly witnessed Pavitt's presentation in Lisbon, which, in any case, was published in December 2000. On 10 April 2001, almost exactly one year after his first remark, Papon (together with Peter Tindemans) authored a statement on behalf of EuroScience, which commented on the process of organizing the sixth edition of the Framework Programme:

> We need a European, 'denationalized' funding policy for the science base so as to compete at the frontier of science, to train people and to build up knowledge bases for the new industries and public priorities of the future. [. . .] One (or more) independent European Research Councils, could play a role similar to that of the National Science Foundation (NSF) or the National Institute for Health (NIH) in the USA.[48]

The potential ERC has remained a complementary scenario to the Framework Programme, but its purpose was no longer to fund infrastructure; taking over Pavitt's line of argument, instead, its justification was now three-fold: European-wide competition, building up the science base, and training people.

The Europolis–EuroScience connection was one of the earliest to promote wholeheartedly the new narrative of an old idea across Europe; as much as academic chitchat was important, however, it needed more to make the step towards formal political recognition. That this task was accomplished was primarily the result of the efforts of a group of distinguished research managers in Scandinavia. A handful of Swedes, in particular, played a crucial role. As it happened, their country hosted the EU Presidency in the first half of 2001. Even though the Swedish government would not officially support the ERC during its term,[49] the country saw a series of meetings related to European research policy, and the Swedish advocates successfully used those occasions to make their case. Two events were of particular importance (see figure 2.1)

The workshop at the Krusenberg Manor held on 25 April in Stockholm gathered some thirty participants under the topic 'Cooperation and Competition – striking the balance in R&D policy'. Participants vividly remember that a future ERC had been the main topic on and off the floor; the minutes, however, didn't yet dare to record things as they were. The 'tension between cooperative and competitive modes of strategy in R&D policy' was 'illuminated

Date	Title (dub title)	Place	Organized (funded)
5.-6.6.00	Europolis workshop	Lisbon, PT	EuroScience (EC)
26.-27.02.01	Europe with a human face	Uppsala, SE	RJ (EC)
25.04.01	Cooperation and Competition-striking the balance in R&D policy (Krusenberg workshop)	Stockholm, SE	Swedish Research Council (SE Presidency)
23.04.02	A new science policy for EU (CNERP conference)	Stockholm, SE	CNERP (RJ)
07.-08.10.02	Towards a European Research Area 'Do we need a European Research Council?' (Copenhagen Conference)	Copenhagen, DK	Danish Research Council (DK Presidency)
19.02.03	Life Sciences in the European Research Council. The scientists' opinion (1st UNESCO Conference)	Paris, FR	ELSF
28.-29.05.03	Life Sciences in the European Research Council. Concrete proposals concerning grants, infrastructures and delivery mechanisms (Venice Conference)	Venice, IT	ELSF
21.-22.10.03	Interdisciplinary meeting – A European Research Council for all sciences (Dublin Conference)	Dublin, IR	ELSF
16.-17.02.04	Europe's search for excellence in basic research (Dublin Symposium)	Dublin, IR	(IR Presidency)

Figure 2.1 List of consecutive events with relevance to establishing ERC

Compiled by the author.

Date	Title (dub title)	Place	Organized (funded)
23.–24.02.04	European Research Council – an Initiative for Science in Europe (European Parliament Meeting)	Brussels, BE	ISE
1.3. 04	Changes and Challenges for European Research Structures and Politics (Harnack House Meeting)	Berlin, DE	MPG
25.–26.10.04	Making reality of the ERC-A novel approach to science policy-making (2nd UNESCO Conference)	Paris, FR	ISE
19.9.05	High-level 'round table' on the scientific autonomy and self governance of the European Research Council (Renaissance Hotel)	Brussels, BE	EUROHORCs (EP)
18.–19.10.05	1st ERC Scientific Council Meeting	Brussels, BE	ERC (EC)
09.–10.11.05	Celebrating the first concrete steps towards the implementation of the ERC (3rd UNESCO Conference)	Paris, FR	ISE

Figure 2.1 (continued)

[. . .] from a great number of angles', they state vaguely. Later, Enric Banda, Secretary General of ESF, was quoted as saying that 'there is in fact not sufficient competition at the European level'. Summing up the meeting, Hans Wigzell, director of the Karolinska Institutet in Stockholm and influential adviser to the Swedish government, raised the hypothetical question of whether 'the situation called for a revolution or just strong improvements in the present European system'. And, a few sentences later, the authors of the minutes added that 'there is a lack of competition on the European scale in basic science. Maybe it is time for a European Research Council in some form?'[50]

Since the workshop was an event in the official calendar of

Sweden's EU Presidency, mentioning an 'ERC' in the summary of this meeting implied that it was now also formally incorporated into the vast corpus of EU documents. The significance of this short notice should not be overstated. More important was the fact that the mathematician Mogens Flensted-Jensen, then Vice President of the Danish Research Council, was among the participants. Agitated by his Swedish colleagues, Flensted-Jensen returned to Copenhagen with a mission. Denmark would take over the EU Presidency in the second half of 2002, and he would make sure that it would dedicate a Presidency Conference entirely to an ERC.[51]

The distance of more than one year between the Swedish conferences and the Danish presidency was convenient, on the one hand, as it allowed a thorough preparation for the event. But it also posed a problem to the ERC advocates: How to keep the ball rolling, in order to spread the message further among colleagues and to start working on policy-makers? Being on the trail from one conference to another was clearly not sufficient. What was needed for the campaign was a dedicated infrastructure. At another conference, entitled 'Europe with a human face', the Swedish delegates found the solution. The meeting took place even before the Krusenberg workshop, on 26–7 February 2001 at Uppsala University's Gustavianum. Formally summoned by the Commission as part of a series of gatherings to present research funded under the fifth edition of the FP format's socio-economic programme and 'to stress the crucial role of the social and human sciences for Europe,'[52] its discussions revolved once more around 'the need for a reorientation of current research policy of the European Union'.[53] One of the Swedish convenors of the conference, Dan Brändström, later noted,

> because of the markedly critical attitude towards the current direction of EU research policy, this conference has become the starting point for the establishment of the Committee for a New European Research Policy (CNERP). The Committee consists of representatives from research councils, academies and foundations and will aim [. . .] at strengthening EU support for basic research efforts in excellent environments.[54]

Under the chairmanship of Brändström (who, as CEO of the Riksbankens Jubileumsfond (RJ) also provided the financial means to cover costs), CNERP brought together Michael Sohlman, Managing Director of the Nobel Foundation, Gunnar Öquist from Umeå University, a member of the EU Research Advisory Board (EURAB), Bertil Andersson, Vice Chancellor of Linköping University, Uno

Lindberg and Olle Edqvist from the Royal Swedish Academy of Sciences, among others.[55] The committee probably started working in earnest in the autumn of 2001, and, despite only being a temporary and rather informal institutional setup, it proved to be crucial for the entire chain of events that would follow. But why was this initiative taking place in Sweden and not, say, in one of the other countries that hosted the EU presidency in the aftermath of the Lisbon Strategy, such as France, Belgium, or Spain? Or, to put it differently: Why was there a group of high-ranking members of Swedish academia convinced and committed to persisting with an idea that its own government was not willing to support?

Partly, the commitment of this group derived from a shared sense of mission that Sweden, where research policy had always had a special place and where the Nobel Foundation was only the most distinguished of several thousand other organizations dedicated to funding and rewarding scientific research,[56] should take the lead in overhauling the European research landscape. The Swedish felt particularly close to the case of academic research, and their frustration with the way the Union handled research policy was contradicting their strong feelings about European integration. There was, however, also a more practical side to it. Over the past decade, Sweden had completely overhauled its public research system, with the consequence that more emphasis was put on competitiveness.[57] Those involved in the restructuring process saw with some concern that resources for funding academic research at national level were dwindling. Couldn't Europe augment (or maybe even completely replace) the nation states in that respect? Since they believed to have a very intimate knowledge of how to achieve their goal, they made their case with a great amount of confidence. 'The EU has to have a vision', as one of the group's members, Pär Omling, head of the newly created Swedish Research Council (Vetenskapsrådet) stated in early summer of 2001, adding assertively that he supported 'the idea of a European Research Council, run by researchers and financed by the European Commission.'[58]

2.3 Everybody is talking about something different

That was an ambitious statement, but Omling did not reveal why the Commission should embark on such an adventure in the first place. Thus, the issue remained uncharted for the moment. Instead, the conferences in 2002 hammered away at the core message: to compete with the US, and to reach the Lisbon goals, 'basic research must be funded by the EU'.[59] The culmination of that rhetoric was reached at the Copenhagen conference in October 2002. In many respects, this meeting was intended to get the elite of the European science community attuned. In the keynote lecture, Keith Pavitt presented his 'radical proposal'. The conference report opened with the statement that 'a clear majority of the interventions [. . .] found that the time had come to begin setting up an ERC'.[60]

Such reassurance was necessary, in part, because, all through 2002, there had been doubters and deniers alike. Norbert Kroó, for example, leading member of EuroScience and vocal representative of the Eastern European accession states, endorsed the idea of course, but expressed doubts about whether it would become a reality within the next ten years.[61] Others were distinctively more critical. Invited by *Science Magazine* to elaborate on the idea, George Radda from the UK Medical Research Council asked: 'Is an administrative structure like the ERC truly necessary?'[62] His answer was a clear 'no'. It was hindering European coordination and, in any case, 'everybody is talking about something different', as he complained in a round-table talk months later.[63]

With the latter remark, at least, Radda had a point. The four questions around which most of the discussions revolved between late 2001 and summer 2003 were: who should provide a budget for it? Who should be put in charge of setting it up? Who should be tasked to administer it? And what should be funded? The ERC advocates had little more to offer than the hope that, with the Lisbon Strategy's ambitious goal of 3% of GDP for R&D, 'fresh money' would eventually become available for 'basic research' at European level.[64] Many deemed the most promising strategy to be to intervene in the on-going 'Convention on the future of Europe' under former French President Valéry Giscard d'Estaing by putting academic research firmly and explicitly into the draft of the new EU constitution.[65] That would then eventually oblige the member states to set up a research-funding agency at European level. But as much as this was a long-term strategy, it left many questions open. In the meantime, the ERC idea inspired more and more circles of the scientific community; different imaginations about its purpose and shape came into play, and people started looking for other solutions. From all the noise that a future ERC created in this period, three attempts to shortcut the long-term strategy can be distinguished, and it is worth taking a closer look at them, not least because they indicate how ambiguous the ERC idea still was at its core. Early on, Ernst-Ludwig Winnacker, a self-declared

'ERC enthusiast',[66] suggested in January 2002 that an ERC should be centred 'at the independent, non-governmental level [. . .], in parallel to the governmental level of the European Commission in Brussels' by proposing an expansion of 'cross-border cooperation' between existing national funding agencies.[67] At the same time, Enric Banda argued that the European Science Foundation (ESF) could administer an ERC, provided that '[n]ew internal structures' were 'put into place to cope with added responsibility for funding.'[68] Both Winnacker and Banda proposed furthermore to use existing funding streams (and in particular the Eurocores programme managed by ESF) as a nucleus of a future ERC.[69] Some months later, an un-named author in *Nature* proposed that, given the conceivable difficulty of compiling sufficient amounts of funding to cover all research fields right from the start, 'one of several possible pilot schemes' could be to implement 'several high-profile European research awards in the life sciences.'[70]

All aforementioned suggestions offered practical solutions to the obstacles ahead that the introduction of an independent, new research funding instrument would face. Winnacker offered the view that existing sources from national funders could be taken for stocking an ERC instead of foundering political negotiations – a shortcut to foreseeably lengthy European political procedures with their uncertain results. Banda offered that the ESF could be tasked with taking over the administrative work instead of building up such an institution within the contested European administrative space from scratch. The Eurocores scheme was regarded as a template for funding schemes at European level. And the unknown *Nature* author proposed that the life sciences, as the research domain with the highest degree of European integration, would be quickest at putting an ERC in place; other disciplines could follow up later.

However, those advances were met with some scepticism. For one thing (and hardly surprisingly), each of them was also advanced out of self-interest. As President of the Deutsche Forschungsgemeinschaft (DFG), one of the oldest and most potent funding agencies in Europe, Winnacker was also chair of Eurohorcs, an informal meeting of the heads of European national funding agencies. Naturally, the national funders regarded the debate of an ERC with suspicion, and bringing it under their wings would put oil on troubled waters. Banda was the Secretary General of the ESF. That organization had attempted a major reform in the 1990s,[71] but because of the inertia of its more than sixty national member organizations, it had remained an ambitious organization with weak resources. The ERC would offer new possibilities in this regard.[72] As for the anonymous author in *Nature*, biology was proudly declaring its leading position among scientific disciplines.[73] Ambitious efforts to create a truly European life science, however, had recently seen unexpected setbacks, as the European Commission had cut funding for two major infrastructure projects.[74] A European research council would certainly have alleviated those pains.[75]

There were also good reasons that argued against each of the advances. The proposal of putting an ERC on top of national funding agencies had two problems. First, even if there were unanimous readiness within the Eurohorcs to contribute a fixed part of their annual budget to a European pot that would then be distributed without being earmarked on a national basis (which was questionable), it was still doubtful whether each of the national funders would have the political and legal remit to do so.[76] Clarifying this issue would have meant involvement in each country's domestic political scene and would, consequently, have scattered the political efforts of the academic community – instead of one joint European effort, it would have been fifteen national ones. Second, the overall budget for academic research in Europe would not increase, at least in the beginning. Winnacker's own suggestion – € 25 million per annum[77] – was tiny, and it was paid out of the existing budgets of the national funders.

As for the idea of putting the ESF in charge of the ERC, after the Stockholm meeting Enric Banda had quickly initiated a working group tasked to 'review the option of creating a European Research Council.' Not surprisingly, one of its conclusions was that establishing an ERC 'is likely to be achieved most effectively through use of an already legally constituted body with the appropriate culture, characteristics, and stakeholders, such as ESF.'[78] The problem was that, as even that work group acknowledged, in its current form the ESF would not be able to carry the ERC. It 'would require a complete reappraisal of its role and mode of operation'.[79] Along the same line, the Copenhagen conference report emphasized that 'ESF must become more effective by being less consensual'.[80] Given the experience of the past, who guaranteed that another attempt of reorganization would be more successful than the last one?[81]

Then to the question whether Eurocores and similar collaborative European funding schemes could be used as basis for a future ERC. The problem with those schemes was that, while they were announced at European level, they still required negotiating the ultimate decision with the national funders.[82] Furthermore, the idea to pool existing schemes opened the question of ownership – bringing Eurocores (conducted by ESF) and programmes conducted by the Commission under FP6 under the same roof[83] would have required a complicated set of negotiation rules between the Commission, ESF, and national funders. The ERC, in this scenario, would become a clearinghouse more than an independent decision-making body.[84]

With the European Life Sciences Forum, the life sciences were probably better organized than any other group of researchers in Europe. Their advance – 'an ERC for Life Sciences' – was persuasive enough to make it into one of the 'six illustrative scenarios' in the discussion paper distributed in advance of the Copenhagen conference.[85] According to anecdotal reports, the idea was intensively debated during the conference. The majority of participants felt strongly that an avant-garde position of the life sciences would split the academic community and raise envy among other disciplines; the summary report noted of 'warnings against the fragmentation of efforts towards an ERC.'[86] While some of its representatives continued to argue that the 'ERC should start small and from scratch, using the expertise of existing bodies',[87] the ELSF adopted the notion of 'an entity that covers all areas of research'.[88]

In summary, there were well-founded reservations against each of the four shortcuts proposed by different groups. Still, they remained viable options, because, while the abstract idea of an ERC had become so powerful, its fully fledged realization seemed increasingly improbable. The Copenhagen event had brought to the fore many different approaches and understandings. In their report, all the conference convenors could come up with were some preliminary directions. Funding should come from 'fresh money', an ERC should 'use existing EU, intergovernmental, national and other European resources and structures', it should be 'accountable to its funders, but autonomous in its operations and run by highly respected scientists', and it should rely on a 'rigorous and transparent peer review process' in order to 'cover all fields of science'. Even more notably, the core question still remained in the title of the report: 'Do we need a European Research Council?'[89]

Notwithstanding the continued lack of clarity, the scientific community was more obsessed with the ERC idea then ever. A number of statements followed swiftly the Copenhagen conference, such as the organization of all European academies (ALLEA); so started a series of conferences in order to elaborate further 'the scientists' perspective' in February 2003; ESF published its working group paper in April; Eurohorcs announced a (rather inconclusive) 'declaration on reinforced research cooperation in Europe' in May.[90] Even EURAB, consisting half of scientists and half of representatives from European industries, made a positive statement.[91] The reason for all that fuss was, partly, to raise political awareness. But, reading those statements and reports today, it is difficult not to gain the impression that they were also painful attempts to come up with some sort of minimal consensus. It was ambiguous and very abstract: A complementary instrument to the FP format, the ERC would be steered independently by a scientific senate, based on a mixture of re-allocated funds and fresh money, and, while legally a stand-alone structure, still somehow drawing on existing structures, such as ESF or the European Molecular Biology Organization (EMBO).

No wonder, then, that frustration was growing among the most fervent advocates of an ERC. David Grønbæk from the Danish Research Councils who had been involved in the Copenhagen conference and its summary report and who knew the different positions of the various players very well, published an article in which he expressed his impatience. Grønbæk framed the discussion of the past months as a conflict between 'incrementalists' and 'radicals'. While the latter contended 'that an ERC should not be rooted in any existing research organization but be directly accountable to the European political system and guided by the scientific community', the former thought of it as 'a lightweight co-ordinating body with little independence, a kind of "agency of agencies" controlled by national research councils, or based on existing European research organizations'.[92]

Grønbæk accused the Commission, national funders and existing research organizations alike of watering down the scope and resources of the future ERC because of their 'institutional interests'.[93] The problem he and his fellow 'radicals' were facing, however, was that, although the 'ERC has now established itself on the European research policy agenda much more firmly than before,' admittedly it was lacking 'a realistic vision.'[94] Indeed, during the first half of 2003 it had become clear that the attempt to induce academic research in the new EU treaty was failing, and the prospect of increased public funding based on the Lisbon Strategy was turning out to be a political hoax. If the very premises on which the ERC debate had started had no political value anymore, how would it be able to create a new, resourceful, and autonomous institution that Grønbæk was envisioning? The radical proposal was on the verge of collapse.[95]

— 3 —

EUROPEAN VALUE ADDED

In Stockholm in April 2002, participants of the CNERP conference had decided 'to lobby EU ministers to endow ERC by taking a tithe from Framework and other programmes'.[1] That was a vague recognition of the political machinery into which the ERC idea would have to be lifted, and whose treatment it would have to survive, in order to make the ERC a reality. Putting the topic on the Council of Ministers meeting minutes was the logical first step, and achieved through the Copenhagen meeting (which is why this conference really was a decisive moment in the pre-history of the ERC). At the next meeting of the Council of Ministers after the Copenhagen conference, presided over by the Danish minister for research Helge Sander, it was concluded

> to continue discussions on a concrete basis, on the purpose and scope of a European Research Council and to explore options for its possible creation, in cooperation with relevant national and European research organizations.[2]

And then? It is a telling anecdote of the provisional character of the ERC campaign that none of the conference participants had thought about that.

One serious attempt to overcome looming inertia was to continue the organization of conferences dedicated to the topic; this was now taken over by the ELSF and, later, by the Initiative for Science in Europe (ISE).[3] The latter was in many respects a new bottle for old wine, since it was driven mostly by the same people already running ELSF – most notably, Frank Gannon, Director of EMBO, and Luc van Dyck. In any case, the events would help to significantly broaden the scope of disciplines and institutions across the academic spectrum, and to unite their claim for an ERC. Furthermore, and not least because of this extended reach, the various national governments could be lobbied even more intensively through informal connections of its individual members flocking around these, such as former Portuguese minister José Mariano Gago (who would become the first president of ISE). However, while it may have been gaining a foothold in a few national ministries, the perplexing question remained how to get the ERC idea through the complex political procedures of the EU polity without being killed by one of its many adversaries.

To that end, the other (complementary, rather than counteracting) approach was decidedly more target-oriented. It arose from a plan frantically devised by Mogens Flensted-Jensen, Dan Brändström, Julio Celis, and others around the Copenhagen Conference in late December 2002, when they realized that a mere note in the minutes of the Competitiveness Council would not suffice to keep the ERC-idea going. Neither would the European Commission follow up on it, nor was the ERC on the priority list of the governments hosting the presidency in 2003.[4] To build up leverage, more was required. Only a few days were left of the Danish EU Presidency, when Minister Sander sent off an official letter. In it, he informed his colleagues from the Council that, based on the conclusions in November, he had invited 'Federico Mayor, former Director General of UNESCO, to chair a small expert group with a year's time [to] present possible options for creating an ERC.'[5] And so, the European Research Council Expert Group (ERCEG), or Mayor Group, was established.

3.1 Clear ownership

Establishing ERCEG was an improvised move: Celis, a research professor at the Danish Cancer Society in Copenhagen and secretary general of FEBS, had convinced Mayor to lend the cause a prominent face; Brändström supplied funds from the Swedish RJ.[6] Despite the haste in setting it up, the group would turn out to be the decisive instrument for the further negotiation on the ERC. For one, it was distinctively different to the ESF working group that was also still in session. Not only were its members hand-picked by Flensted-Jensen, Brändström, and Celis, and thus firmly committed to the idea of a financially strong and politically independent ERC.[7] Because of its political assignment, ERCEG basically extended the political remit of exploring the ERC for another year. Thus, the group quickly gained the ultimate authority within the scientific community, and also consolidated the lobbying efforts previously accomplished by CNERP and now continued by ELSF, by making its case to high-ranking politicians in different European countries.

In its introductory letter to the ERCEG report, Federico Mayor would later list the discussions with different stakeholders in the process, such as EURAB, ALLEA, EuroScience, ELSF, and many more.[8] However, the most significant negotiations during ERCEG's sessions must have been with the European Commission, which was represented at ERCEG meetings by an observer. Indeed, it seems as if the Commission leadership used the Mayor Group to tactfully ring the changes of its own approach towards the ERC, without having to fear that this would prematurely leak to the public and without having to compromise with any other stakeholder (such as the ESF that was still eager to play an important role in setting up the ERC).

Bringing the Commission on board the campaign was crucial in order to keep the ERC idea in the political realm. It was the Commission, or, to be more precise, the team around the Director General for research, Achilleas Mitsos, that developed the core arguments that would, eventually, convince the politicians across Europe of this idea's relevance. Under Mitsos, the ERC campaign, which so far had been mostly a grass-roots movement of highly engaged scientists and science managers, became a well thought-through theatre staged by a group of shrewd high-ranking functionaries in the Commission's Directorate General for research. Thus, the Commission was not simply joining forces; in some respect at least, it was taking over the campaign. The place where this was negotiated was in the ERCEG.

From the outset of the ERC frenzy, the Commission had been sitting on the front row. It had actually sponsored the first meetings where the new narrative was presented; its representatives were at the Krusenberg Manor meeting and probably also at the CNERP meeting in April 2002; Commissioner Busquin personally attended the Copenhagen conference as well as at the ELSF conference in Paris in February 2003. In all those meetings, however, and to the dismay of the advocates (who later accused the Commission of being 'hostile' towards the ERC)[9], the Commissioner and his emissaries had carefully repeated the same cautious message consisting of the following three parts: the ERC 'would fit very nicely into the European Research Area project'; since it was an idea created by the scientific community, the national funders or an existing European research organization should be tasked to set up the ERC; and, anyway, there was no budget in the current edition of the Framework Programme available for providing funds for such an organization.[10] This line was also taken in the Commission Communication where the ERC was explicitly mentioned for the first time.[11]

The reason for this cautionary approach was two-fold. For one, in many encounters with the advocates of the ERC early on, Commission representatives must have perceived the ERC idea as an assault on their FP format. Indeed, loathing the Framework Programme may have been the one feature that unified the advocates with their otherwise so different conceptions of the ERC. It was not helpful to establish lines of communication. Even though its rank and file was probably more attached to the various programmes within the FP format and thus may have felt assaulted by the rhetoric of the ERC advocates, the leadership around Busquin and Director General Achilleas Mitsos remained receptive to what was going on. After all, the Commission itself had made several attempts to change the trajectory of European R&D policy; ERA had only been the latest one.

During the first months of 2003, while the ERC advocates were growing frustrated and ERCEG was still being set up, Busquin and Mitsos must have re-evaluated the ERC and now found it a political project worth supporting. Several reasons might have fed this change of mind. For one, the Commission came to realize that the implementation of ERA would be considerably slower than initially anticipated. Even with a new policy architecture clearly prioritizing R&D, overcoming the inertia of its traditional means turned out difficult.[12] Furthermore, with the sixth edition of the Framework Programme up and running, it was already time to start thinking about the next one. Weighing up the pros and cons of supporting the ERC idea, Busquin and Mitsos, with their trusted advisers (Peter Kind, Jean-Eric Paquet, Robert-Jan Smits, and others), now came to the conclusion that there was little to lose, but much to win. At worst, endorsing the ERC would lead to a rejection through the member states; at best, it would substantially increase the standing of the Commission among the scientific community, and significantly raise its leverage on national R&D as well as its monetary heft with the next edition of the Framework Programme – not to mention the potential impact that it might exert on the shape of the FP format itself.[13]

The ERCEG meetings turned out to be the ideal setting for discreetly negotiating with leading ERC advocates the basic conditions under which the European Commission would, in principle, join the ERC campaign, and carry it through the intricate political process. For the ERC advocates, these discussions also entailed facing some bitter truths. First of all, as the Mayor Report acknowledged, the EU Treaty stated that any European funding for research 'will have to be via a specific item in the budget for the EU Framework Programme'.[14] The authors of the report did not any longer cherish illusions about the fact that the ERC would have to be under the wings of the Commission. Maybe even worse, when the report turned to the ERC's 'full operational autonomy in all scientific and scholarly matters including funding policies and decisions', another key demand of the advocates, the conclusion remained ambiguous, acknowledging only that '[t]he legal framework for the ERC will ultimately depend on the outcome of negotiations between the Member States, the European Parliament, and the Commission.'[15]

Still, ERC advocates within the ERCEG as well as outside embraced the fact that the Commission was finally moving in their direction. After all the uncertainty, a powerful political institution had taken 'clear ownership', just as had been demanded since Copenhagen.[16] And, to alleviate fears of the Commission taking charge, the report introduced the idea of a 'European Fund for Research Excellence': 'In this way it is possible to establish the ERC with reference to the EU budget, while keeping its management at arms-length from the Commission.'[17] It was an elegant formulation, but, as we shall see later, it could allay concerns only temporarily: what, exactly, did 'at arms-length' mean? For the moment, however, positive reactions prevailed.

By the time the ERCEG was about to conclude its work, the ERC idea was also gaining momentum in the political realm, mostly thanks to the organized efforts around the two ELSF conferences in the first half of 2003 (see Figure 2.1). Only in autumn 2003, Commissioner Busquin publicly announced his support for the ERC campaign.[18] There were good reasons to remain cautious. In another meeting between the Mayor Group and representatives of the member states in October 2003, Busquin put the essential question on the table: 'are the members [the EU member states, TK] really willing to give more to fundamental research?'[19] The answer to that question would have to be (almost) unanimously 'yes'; that meant that advocates and Commission jointly had to align the science ministries of more than two dozen sovereign governments. It was advantageous that the two member states taking over the EU presidency for 2004 each put the ERC on their agenda.[20] But as a prerequisite to all that, a convincing argument had to be elaborated why the ERC, as a new political instrument for research funding, would be required at European level at all.

3.2 Lack of sufficient competition

To put an idea such as the ERC into reality, the sequence of formal steps, as stipulated by the multi-level procedures of the EU polity, has to be carefully observed.[21] The European decision-making procedures are compartmentalized in different cycles (interacting through complicated mechanisms), with the half-year cycle of EU presidencies complemented by the multi-annual cycle of programmes (such as the latest edition of the Framework Programme then running over five years), and the even longer cycle of budgetary planning (the so-called multi-annual financial framework (MFF) laying out over seven years the financial resources of the Union programmes, including the budget of the European Commission).

By late 2003, time was already pressing: negotiations on the next edition of the Framework Programme would soon be in full swing, as would the intricate process of establishing the next MFF. With combined efforts, the ERC advocates had created enough leverage to push the ERC onto the political stage: in September 2003, the European Parliament expressed its wish to include the creation of an ERC in its discussion of the next (seventh) edition of the Framework Programme;[22] around the same time, at a Competitiveness Council meeting, the French delegation proposed to ask the Commission for a Communication explicitly on the topic of basic research funding.[23] That instigated a series of formal exchanges, each of which the team around Mitsos used to entrench the ERC idea a little more deeply in the discursive murmuring of the European institutions.

For this purpose, Mitsos drew on an argument probably developed during the ERCEG sessions. Despite its otherwise rather vague provisions, the Mayor Report featured two crucial refinements. The first was that 'the ERC should during the first 3–5 years reach a grant volume of at least € 2 billion a year,' and that the fund 'should be a specific item in the budget for the next EU Framework Programme.' The other was that, as 'an essential part of a new, forward-looking definition of European added value', the ERC should specifically 'fund research teams of the highest quality, regardless of their national location, and chosen by international peer review of applications'.[24] Probably it was the single-digit number that attracted the ERCEG members; politically, the more important provision was the narrowing down of the future ERC's funding options.

At the conferences during that time, calculations about how much funding academic research at EU level would actually cost were circulating in abundance; in Paris, for example, the most common number mentioned was € 1 billion per annum.[25] But those were castles in the air. ERCEG, on the other hand, was negotiating with the Commission; and it was probably Achilleas Mitsos who determined the exact number for which to ask.[26] Similarly, calibrating the ERC on funding competing researchers across Europe was certainly not against the will of the ERCEG members (as similar ideas had been around). But in a speech at the third follow-up conference organized by ELSF in Dublin in October 2003 – two months before the Mayor Report was published – Mitsos used almost exactly the same phraseology as the report would, declaring that 'now it is time to bring a new definition to European Value Added.'[27] And building on exactly the same argument that the report would make, Mitsos meant by that 'the principle of allowing a researcher in any one of our member states to compete with all other researchers to win funding'.[28]

It is thus not an exaggeration to say that Mitsos and his team instilled two refinements in the Mayor Report that they felt they would need for bringing the ERC campaign through the EU decision-making procedures. That only underlines again how useful the ERCEG sessions had been for the Commission to quietly prepare joining the ERC campaign and to attune with the ERC advocates. But what did the two specific provisions mean? As for the number, it was simply a tool for negotiating a higher budget for the next edition of the Framework Programme. The focus on individual researchers, however, meant nothing else than completely overturning the rationale that, until now, had determined European research policy. The lever to do so would be the 'added value' of European research policy.

In the European R&D policy discourse, the notion of 'added value' was the specific equivalent to the more general principle of subsidiarity that had been an increasingly important catchword in Union politics since the Maastricht Treaty; like subsidiarity, it dictated 'that the Union does better than the Member States'.[29] However, and again like subsidiarity, added value was a rather opaque concept; it could be used by member states to protect their national interests as much as by the Commission 'to increase activities in previously excluded areas of policy'.[30] So far, the Framework Programme's main political goals were framed as enabling transnational research cooperation in, and as enhancing mobility across, states; effects of funding that could only be achieved at European level. That did not exclude 'basic' research – many consortia consisted of partners from universities, and some (smaller) instruments were specifically dedicated to them.

However, a separate mechanism directed at academic research would also require a separate though complementary rationale (see also Figure 1.1). That's why, when the Commission published its Communication on 'Europe and Basic Research' in response to the Council's assignment, competition was put centre stage. In style and content, the paper resembled previous reports on R&D, except that the focus was not on overcoming the 'European paradox', but on assessing the science base. Despite its many strengths, the Communication concluded, 'the lack of sufficient competition at European level' offered 'a less attractive environment for researchers' and thus posed an essential weakness.[31] But relief was possible:

> By stimulating competition and encouraging innovation as well as experimentation in ideas and new approaches, including interdisciplinary ones, it [a new support mechanism, TK] would stimulate creativity, excellence and innovation by exploiting a form of European added value other than that produced by cooperation and networking: the added value which comes from competition at EU level.[32]

And what was the essential means to stimulate the desired competition? A 'European level support mechanism for individual teams' research projects, modelled on the "individual grants" given by the NSF.' It was, of course, exactly the same idea as in the Mayor Report (to which the Communication dutifully referred),[33] only by now it was tidied up.

And it worked: the new rationale resonated within most member states' governments.[34] Actually, only three governments put up resistance. The first to voice resentment was the Finnish representative; already at the informal meeting with the Mayor Group, Sakari Immonen had stated that he was 'radically critical' of the idea.[35] Strong reservations were soon signalled also from across the Channel. But when Lord Sainsbury, Minister for Science and Innovation, was scheduled for a critical speech at the Irish EU Presidency Conference on the ERC in Dublin on 21 February 2004, he was urged to read the Mayor Report. Sainsbury was convinced of the sincerity of the ERC idea. Since he couldn't change his prepared speech manuscript at short notice, he added a personal note that he was 'greatly impressed by the work which has been done,' and stated:

> If it is going to be independent, focused solely on scientific excellence, and based on a transparent peer review process, we would see it as a valuable mechanism for allocating money to basic research as part of the next Framework Programme.[36]

Most advocates remember Sainsbury's speech as the decisive moment in the ERC's history, and there can be no doubt that his move was an act of intellectual honesty. Still, overcoming the British resistance was rather easy. Ever since George Radda's misinformed statement, influential institutions such as the Royal Society had shared the opinion that the EU – and the Commission in specific – could simply not be expected to come up with something as sophisticated as the autonomous structures of the British research councils.[37] As his opposition was not directed against the political argument made by the Commission, but rather fuelled by doubt about its ability to realize it, the minister could be convinced that this assumption was based on a mixture of arrogance and ignorance.[38] Once the advocates got hold of the minister in Finland, opposition was vanishing there, too.

In August 2004, the ERC, still under disguise, appeared (as one of six 'mechanisms') in the Commission's first proposal on the next (seventh) edition of the Framework Programme, specifically dedicated to 'stimulating the creativity of basic research through competition between teams at European level'.[39] By then, the Dutch government had inherited the EU Presidency, and only one party was still in outspoken resistance: the Italian government.[40] Its objection was much more fundamental than the British (and Finnish) one. It attacked the political implications behind the Commission's advance to rebrand the 'added value' of European R&D policy.

The Dutch government made an attempt in early November 2004 to reconcile the opposing parties, inviting leading science policy representatives of the two countries, as well as two emissaries from the European Commission (Robert-Jan Smits and William Cannell) to Florence. The Italian position was not against funding academic research, although it argued that, 'from the Community standpoint', it 'should have an integrated function' together with applied research. In that respect, the Italian delegation held that the ERC would be detrimental to the political purpose of European R&D funding. Thus, they did not buy Mitsos' remodelled 'added value' of European competition, for 'all serious research groups do anyway their best to excel.'[41] Against that allegation, the Commission representatives pointed out that Europe-wide competition among researchers would

> ultimately benefit both the research community and the national funding agencies, enabling the latter to make more informed and better strategic judgements, for example, of where and how to invest in order to maximize the country's research potential. It is important for all parties to know how they compare against their peers in Europe.[42]

Still, Alexander Tenenbaum, then General Director for International Development in the Italian Ministry, went on. Focusing on competition would break an 'iron rule' of research funding thus far, 'namely increasing the overall European strength through collaboration, while reducing the differences among member countries.'[43] Tenenbaum argued that a counter-productive effect was to be expected, as 'European resources would be dispersed in such a way as to increase those differences.'

> It should be obvious [. . .] that, in a competition open to national teams, countries that are at the forefront of scientific and technological research will pump resources from the least favoured.[44]

Opening up to European-wide competition was one of the reasons why the ERC was so attractive to its advocates, and that is also why it had won the support of Busquin, Mitsos, and their team. However, there was little the Commission could argue against the Italian objection. The Florence seminar produced some joint conclusions, and when the Dutch prepared the next Competitiveness Council's conclusions on 'future European policy to support research' in late November, they introduced explicit passages on the ERC; specifically, the text acknowledged both

> the case for funding investigator-driven basic research, with a view to supporting research in Europe so as to achieve the highest levels of excellence and creativity [and] the usefulness of examining the setting up of a new operational mechanism aimed at supporting basic research of world class quality through a system of international peer review.[45]

To be adopted, the conclusions required unanimity. But the Italian minister, Letizia Moratti, kept up her resistance and vetoed the proposed text. Annoyed with Italian intransigence, and knowing that 'a substantial majority of delegations' supported the text, the Dutch decided to issue it anyhow, as 'Presidency conclusions'.[46] Despite not being formally adopted, this cleared the way for the Commission, lifting the disguise and putting an instrument explicitly called 'ERC' in its proposal for a joint decision by the European Parliament and the Council 'concerning the seventh framework programme.'[47]

As much as their resistance had been despised by those speaking in favour of the ERC, Moratti, Tenenbaum and their Italian colleagues had a point: the competition at European level would disadvantage some countries, particularly those not investing heavily in R&D. Why, then, did all the other ministries readily accept the Commission's argument and support the Dutch proposal (Figure 3.1)? One reason

Country	On new mechanism	ERC mentioned	Document
Austria	positive	yes	15242/04
Belgium	positive	no	6270/05
Cyprus	positive	yes	7727/05
Czech Republic	positive	yes	6730/05
Denmark	positive	yes	15254/04
Finland	conditionally positive	no	5957/05
France	positive	yes	7153/05
Germany	positive	yes	15195/1/04 REV 1
Greece	positive	yes	6062/05
Ireland	positive	no	6925/05
Latvia	positive	no	7668/05
Lithuania	positive	yes	6030/05
Polish	positive	yes	6269/05
Sweden	not explicit	yes (in passing)	6066/05
Spain	positive	no	6347/05
UK	positive	yes	15162/04

Figure 3.1 Position of member states, late 2004/early 2005[48]

Compiled by the author.

certainly was constant attention – more than four years into the campaign, the ERC had got to the top of the wish list of scientists across Europe. The Greek government (certainly representing one of the countries with low R&D intensity and hence not to be expected to be one of the main beneficiaries of the ERC) stated frankly the 'overwhelming support of the Greek scientific community to the promotion [. . .] of a new action for basic (investigator driven) research'.[49] Elsewhere, the lobbying efforts of ERC advocates added to the success more directly: this was not only true for Great Britain and Finland, but also for Germany (where Wilhelm Krull had a direct line to Minister Schavan), France (as mentioned before), and for Portugal, Poland and Hungary (with José Mariano Gago, Michal Kleiber and Norbert Kroó, respectively, making use of their political clout). It also seemed as if, for the governments of the new accession states specifically, supporting the ERC became a symbol for proving their political maturity. Yet another motive was that policy-makers and representatives of the scientific community hoped to use the ERC for reform of their domestic research policy. Finally, many may have embarked on a strategy of lukewarm endorsement. For example,

Austria emphasized that it still 'regards Collaborative Research as the core' of the next edition of the Framework Programme.[50]

3.3 Credible to the scientific world

In the meantime, the series of conferences and meetings dedicated to the ERC had continued. Now formally organized by ISE, and again under the attentive stewardship of Frank Gannon and Luc van Dyck, those events brought to the fore and specified the manifold visions of the scientific community. More importantly, however, they helped to further attune the scientific communities to what the Commission deemed to be realistic and feasible; consequently, the breadth of the ERC as a collective imagination narrowed significantly down.[51]

It was tacitly agreed by ERC advocates during the ERCEG meetings that Mitsos' team would take the lead in the political campaign for the ERC. That is not to say that the ERC advocates played no role any more; quite the contrary – their lobbying with ministers was crucial. But they had grasped that, without the Commission's intimate knowledge of the decision-making procedures of the EU polity, there was no chance to get the ERC idea through. The Council of Ministers and the European Parliament now began discussing in earnest both the Commission's proposal of the seventh edition of the Framework Programme (including the ERC), and the budgetary provisions for this programme under the next multi-annual EU budget (the MFF). Both discussions were challenging in their own way. Given the advocates' growing impatience, the proposal for the next edition of the Framework Programme would inevitably include discussions concerning the ERC's implementation, discussions that would, at least partially, put in doubt the Commission's sincerity concerning the ERC's autonomy in the scientific communities.

The budget negotiations, on the other hand, would be a hazardous game mostly beyond Mitsos' control: the MFF was almost arbitrarily decided between wary finance ministers, usually discussing just a few highly aggregated budget lines and paying little attention to specifics such as research funding, which represented only a small slice of one of those budget lines. Probably because it was so volatile, the budget issue was actually easier to accomplish. Pavitt and others in his aftermath had claimed that funding academic research was worthwhile.[52] But it would be good to have a more comprehensive analysis, proving that the ERC would contribute to strengthening the Union's economic competitiveness.

Another expert group was set up, this time consisting of representatives from academia and industry alike. The group published a lengthy report in early 2005, which coined the signature label that the future ERC would be tasked to identify and fund: 'frontier research'. According to the authors, the notion combined the ideas of research 'creating new knowledge and developing new understanding'; being 'an intrinsically risky endeavour'; dissolving 'the traditional distinction between "basic" and "applied" research'; and pursuing 'questions irrespective of established disciplinary boundaries'.[53] Unlike 'basic research', which the Commission and the expert group assumed to appear economically dubious to finance ministers and the like, 'frontier research' could hold up the promise of contributing to the overarching goal of innovation and economic growth.

It well accorded with the Commission's proposal to double the budget for research, essentially asking for € 73 billion over seven years.[54] This would have easily covered additional expenses for the ERC. However, the MFF discussion ended up centring mostly on political bargaining;[55] the situation grew critical, when, in summer 2005, the Luxembourgian EU Presidency proposed to cut by half the budget dedicated to competitiveness (heading 1a, where, among other lines, the research budget was located).[56] The Commissioner – in the meantime, Janez Potočnik had replaced Busquin – protested, but to no avail.[57] In December, under the British EU Presidency, the financial perspectives for the MFF 2007–13 were finally adopted. In it was an unusual provision, specifying that, from heading 1a in the budget table,

> EU funding for research should [. . .] be increased such that by 2013 the resources available are around 75% higher in real terms than in 2006. This research effort, as reflected principally through the 7th Framework Programme, has to be based on excellence while ensuring balanced access for all Member States.[58]

It is not entirely clear who had secured this earmark,[59] but it ensured that there would be enough money – namely more than € 7.5 billion – in the budget for the seventh edition of the Framework Programme to endow the ERC. The remarkable success indicated that the ERC campaign had made itself heard in the highest echelons of European policy-makers – ultimate proof that the informal contract between the DG leadership and the ERC advocates had been fruitful.

Mitsos' strategy to secure political commitment for fresh money at EU level – the €2 billion per annum in the Mayor Report as a starter for negotiations – was successful because his remodelled concept of 'added value' was broadly supported by ERC advocates. While everybody could agree that the future ERC should be 'science-led, politically accountable, operationally independent',[60] the answer to where in the European administrative landscape the ERC would fit in order to stay autonomous was much more difficult. The Mayor Report had detailed four options for setting up the ERC: as an organization in one of the member states, as an 'intergovernmental organization' (based on a separate Memorandum of Understanding of all member states), as an 'interagency body' (based on a provision in the Treaty which would later be referred to as 'article 171'), or as an 'executive agency' (following an agency template available to the Commission).[61] The former two fell out of discussion quickly,[62] and debate concentrated on the latter two remaining (see Figure 3.2). Each had a serious flaw. Unfortunately, the Commission and the advocates disagreed about which was more problematic.

Denomination	Legal framework	Position
'Executive Agency' (EA) model	Council Regulation 58/2003	Agency of the European Commission
'Article 171' or 'interagency' model	Article 171, Amsterdam Treaty	Independent structure at European level

Figure 3.2 Alternative models of ERC implementation

Compiled by the author.

During the ISE meeting in February 2004, the jurist Armin von Bogdandy elaborated further on the two options on the table. In his presentation, he favoured the 'interagency' model. Its legal provisions were specified in article 171 of the Amsterdam version's EU Treaty (hence the name), according to which 'joint undertakings or any other structure necessary for the efficient execution of Community research, technological development and demonstration programmes' could be set up.[63] That was a tempting, if somewhat loose, provision: despite funding provided by the FP format, the ERC could be founded firmly outside the Commission's reach. Von Bogdandy did not conceal that he was much less impressed by the 'Executive Agency' (EA) model. Its legal foundation was a recent regulation allowing the Commission to 'entrust certain tasks in the management of Community programmes' to an outsourced entity, all the while the Commission retaining sole power over the structure.[64] This model had been intended to outsource labour-intensive activities to a subordinate agency 'to provide better service and efficiency gains' – the latter meaning primarily, but not only, financial savings due to lower salaries.[65] As von Bogdandy put it in his speech,

> it is difficult for me to imagine that such an auxiliary Agency to the Commission will develop into a prestigious and autonomous institution.[66]

Not surprisingly, many ERC advocates began to rally around the interagency option, not least because they perceived the EA model as one that would curtail the ERC's autonomy.[67]

The Commission, in the meantime, came to a different conclusion. Article 171 made no further legal specifications whatsoever as to how the new structure should be set up, and meant that those details would have to be decided in a separate political process involving the member states and the Parliament.[68] In the given political climate, that opened up the possibility that the ERC would effectively be delayed, if not subverted by governments still giving it only lukewarm support. What von Bogdandy perceived as the great advantage of his preferred solution – 'it should not be difficult to give scientists and scientific organizations a big say'[69] – was countered by claiming that the EA model also

> satisfies most requirements: it is a distinct entity, it ensures accountability to the political and budgetary authorities, it should be able to incorporate scientific governance; it is a model that looks like many national research agencies.[70]

The last two points were overstating the Commission's position: the EA option resembled national role models only from a very superficial point of view, and incorporating scientific governance would prove to be difficult. In any case, it would be naive to see the conflict solely in terms of what was in the best interest of the ERC; as usual, it was driven by rational arguments as much as by vested interests only thinly disguised, namely who should own the ERC (or at least have the right to participate in its governance). The 'article 171' model was advanced not only by advocates, but also by European research organizations, such as the Max Planck Gesellschaft (von Bogdandy's host institution), the Royal Society, and Eurohorcs. For these institutional stakeholders, building a new structure from scratch would have allowed them to claim a seat in the new institution's supervisory board, an option that the 'Executive Agency' model would not foresee. The Commission, on the other hand, drawn into safeguarding the ERC campaign through the political procedures, had no interest in sharing the organizational responsibility of the new instrument.

The diverging opinions posed a dilemma for Mitsos in particular: the issue would inevitably be addressed in negotiations for the upcoming edition of the Framework Programme,[71] which was going on in parallel to the budgetary procedure; already, discussions began spreading into the national ministries.[72] To alienate the ERC advocates would be harming the campaign, which depended entirely on its standing within the scientific community. As Mitsos himself publicly pledged:

> We must not give the impression that the Commission is simply using some scientists to improve its own credibility, no, we must create a system [. . .] that is credible to the scientific world.[73]

As with the budgetary discussion, Mitsos prepared flanking measures in order to convince national ministries and other institutional stakeholders of the sincere intentions of the Commission.[74] The DG started a comprehensive consultation process[75] and, together with Eurohorcs, published yet another report on 'the key principles which should govern the implementation of a European basic research initiative'.[76] But by late 2004, Mitsos must have feared that those measures would not suffice, and so he tasked Chris Patten, former governor of Hong Kong and British Commissioner, to head a small identification committee, which would come up with the principles according to which the future steering body of the ERC should be composed, and also to identify the founding members of that body – the Scientific Council.

After consulting various European stakeholder organizations the committee came up with an impressive list of 22 names[77] – cleverly distributed across different European regions, scientific disciplines, and types of academic institutions, and also balancing advocates and newbies to the ERC campaign. Allegedly, when Patten presented his list, Commissioner Potočnik did not even look at the final list when signing it off. That story is probably too good to be true, but it reflects how carefully the Commission spun its own hands-off approach. The Patten Committee's sound methodology of selecting the founding members of the Scientific Council brought the Commission additional credit among the national ministries,[78] and it lured some longstanding ERC advocates into the Commission's camp.

Not even this extraordinary step could contain the conflict; rather, it was further intensified in the autumn of 2005 at the political level when the European Parliament maintained 'that the ERC, following a brief transitional period, should be established pursuant to Article 171'.[79] In order to come to a compromise, an informal standoff took place on 19 September 2005 in the Renaissance Hotel in Brussels. The institutional stakeholders and a considerable number of members of the European Parliament had invited Potočnik and his team. The Commissioner maintained that the EA model 'offers a satisfactory response to the problem of scientific autonomy', it 'constitutes a well-defined regulatory environment', would not distract attention to 'issues not related [. . .] (e.g. location)'; most importantly, Potočnik emphasized that only this model would be able to 'deliver an ERC within a shorter time frame and thus take full advantage of the momentum gathered around the ERC.'[80] The other side weighed in that, while it would be possible

> for the Commission to maintain artificially a functionally autonomous ERC on a temporary basis for a short initial period, [. . .] the body needed to grow up and stand on its own as an independent autonomous and professional body, operating within appropriate broad objectives and financial controls.[81]

Obviously rallied by advocates who were becoming increasingly furious by the insistence of the Commission, the phalanx of countries opposing the Commission was impressive. Yet it was similarly obvious that, at this point in time, only the EA model would make it possible for the ERC to start already in 2007. Hence the Commission dryly calculated that, in the end, momentum would trump, and that setting the ERC up would be deemed more important than specific concerns against the EA model. In preparation for the next Competitiveness Council meeting, the Commission convinced the British EU Presidency to come up with a compromise, stipulating that

> an independent review will also be carried out of the ERC's structures and mechanisms [. . .]. The review will explicitly look at the advantages and disadvantages of a structure based on an Executive Agency and a structure based on Article 171 of the Treaty. On the basis of this review, these structures and mechanisms should be modified as appropriate.[82]

Several ministries only grudgingly agreed, and the Swedish minister refused to give his consent. But, just as with his Italian counterpart one year before, 'a large majority' supporting the compromise effectively overruled him,[83] allowing the Commission to move on with its preferred solution.

The campaign had successfully overcome three major obstacles in the political procedures: the move towards excellence; the financial commitments; and the organizational structure. Now that the ERC was palpable, it was very different from that initially devised by its advocates, in terms of its position as well as its organizational structure. Instead of an instrument replacing (or being an alternative to) the FP format, it was determined to become part of its next edition. How exactly would the ERC be able to prove its uniqueness; along which principles would it distribute its funds; how would it operate? These questions will be addressed later, in Chapters 6 and 7. More imminent, instead of an independent funding agency modelled after the NSF, the ERC was determined to become a legal compound of different legal entities – the freshly set-up ERC Scientific Council, the not yet existing Executive Agency, and the still to be negotiated parental legal framework of the seventh edition of the Framework Programme. How could this ERC compound be brought together in order to function harmoniously?

— 4 —

THE MOST PROMISING OPPORTUNITIES

By the time the Scientific Council gathered for the first time, neither the inception of the ERC, nor the additional budget necessary for extending the funding lines of European research funding were yet approved by the other European institutions, the Council of the European Union and the European Parliament. Thus, appointing the Scientific Council more than a year in advance of the beginning of the ERC was a flanking measure securing those two on-going debates. Not by chance taking place on academic territory rather than in a Commission building, the meeting was to prove that the Commission, if it were tasked to set up the ERC, would stick to its promise that the research-funding body would be autonomous. At the same time, it was also signalling the advocates and institutional stakeholders not to doubt the Commission's determination to set up the ERC in its preferred way.

Ever since the ERC campaign had been initiated, there was unanimity that the new instrument should be autonomous and science-led. The master template of how public funding for academic research should be organized, and the model after which the ERC had been envisaged from the beginning of the campaign was the NSF. Its legal provision was a single public law, which became part of the *US Code*.[1] The act foresaw the establishment of a 'National Science Board' as the agency's supervisory body, and a director as its executive organ. Established for an indefinite period, the NSF has been a stand-alone organization ever since its creation, meaning that, within the sprawling US 'federal executive establishment', it has been a 'non-regulatory' 'commission' 'outside the executive departments' reporting directly to the US President.[2]

In the firm opinion of many advocates, the ERC should have been set up along the same lines: that its autonomy, that is, its deliberate power to design and conduct impartial and scientifically objective decision-making procedures for the allocation of funds, was to be ensured by the instrument's institutional independence. Yet this opinion was driven very much by an anti-Commission impulse, and it was blind to the fact that institutional independence offered no protection from political meddling. Even a decades-old agency like the NSF was regularly exposed to the – sometimes highly detrimental – policies of US Congress;[3] to protect itself, it relied on well-established relationships to policy-makers and discursive routines that proved its value. For a new funding body in the EU's even more complex multi-level governance system, this was not a given, even though the ERC campaign had carefully crafted an alternative model of 'added value'.

Thus, when it came to implementing the ERC, exposing it to political meddling would have been a serious impediment from the start, since an institutionally independent ERC would have had to struggle with political demands of more than two dozen member states (making it difficult to abstain from 'juste retour' policies), and it would not necessarily have had the Commission's backing in the financial negotiations taking place every seven years (which would remain predictably abstract and volatile). In other words, this path would have seriously threatened the ERC's most fundamental claim to allocate funding based only on scientific qualities of research proposals.

Politically, the solution for the 'Executive Agency' model was a solid and pragmatic compromise; under the wings of the European Commission, the future ERC would have the advantage of a multi-annual budget for planning ahead, and would be cushioned into the Commission's R&D policy service, thereby shielded from the member states' vested interests. But it also meant that the ERC came to rest on a formula that had completely revised the NSF equation (see Figure 4.1): the ERC's autonomy would be ensured by political protection through the Commission, but with the cost of institutional dependence. The new equation would have to prove its validity. Furthermore, the Commission had succeeded not because of convincing its opponents on the viability of the new equation, but by emphasizing a circumstantial fact: that only if it were established as an executive agency, could the ERC get started in time. The argument had been complemented by establishing the Scientific Council, a signal that this aka-ordinary body would have full authority to draft the ERC's funding calls and decision-making mechanism.

	Autonomy equation	**Potential weakness**
NSF	Institutional independence ensures decision-making autonomy	Exposed to political meddling
ERC	Protection from political meddling ensures decision-making autonomy	To follow regulations of parental organization (i.e. European Commission)

Figure 4.1 Comparison of NSF and ERC on underlying rationale

Compiled by the author.

4.1 What the ERC will need

There were hundreds of expert committees established in the European polity, but the Scientific Council was designed to be more than an advisory body.[4] Never before had the European Commission delegated responsibility over one of its programmes to 'an organization that is independent and autonomous in all matters scientific'. The appointment of the Scientific Council was a negotiating offer: it handed over responsibility to a group of highly distinguished scientists and scholars, thereby asking to accept the Commission's approach of codifying the organizational setup under which the ERC would be run. 'That is certainly not business-as-usual!', Potočnik's notes from the October 2005 gathering continue, yet he later lectures the audience that 'spending public money creates obligations'.[5]

Legally and financially, the ERC would be part of the Commission's own research funding instrument, namely the seventh edition of the Framework Programme, which was destined to start in 2007; it would be autonomous with regards to the strategy of how to distribute funds, but it would be part of the Commission's overall research policy and, thus, also be held accountable to the legal and financial standards of the Commission. Accountability, of course, was of great concern for a supranational bureaucracy that was just resolving a severe crisis of legitimacy from a few years ago through even tighter control measures.[6] That the Commissioner felt obliged to inform his audience again and again that 'one should not spend public money without accountability', only indicates the risk that the Commission leadership felt it had taken. It also put into perspective Potočnik's proud announcement that the Scientific Council's autonomy would be 'a token of the trust we place in the European scientific community.'[7]

Despite the extraordinary remit that the Commission promised, the legal construction of the ERC would imply that its Scientific Council would have no insight into the day-to-day operations of the ERC's administrative branch (the Executive Agency that still had to be implemented). Also, its members would act solely on a voluntary basis and with no resources at their disposal. Thus, the new body was not a governing board: not a board, as it had neither formal decision-making power nor supervisory functions, and not governing, as its remits were restricted to matters of scientific strategy. Instead, it was the 'Scientific Council' of the 'European Research Council' – a strange denomination, but quite correctly pointing to its somewhat detached role from the other emerging entities in the ERC compound.

Why, then, had the highly distinguished scientists accepted the Commission offer? Besides the personal ambitions and duties of the individuals, the answer was two-fold. First, the Scientific Council offered a new opportunity to get practical. Up until now, the scientific community had either been organized through conferences that were open to anyone who advocated the ERC (or wanted to know what it was all about), with sometimes more than eighty participants and widely differing, or through ad hoc committees with short lifespan and limited remit. As valuable as it may have been to the ERC campaign, all that CNERP, ERCEG, ELSF, ISE and other initiatives had contributed to the creation of the ERC was limited to deliberative suggestions. The Scientific Council, on the other hand, was a closed club of twenty-two handpicked members with a tenure of four years (they would be routinely replaced by other representatives from academia). It was a new player with broad yet so far undefined authority and the power to actually implement. Whatever the group would decide, it would be based on its mandate to steer the ERC.[8]

The second reason was timing. As the minutes of their first meeting reveal, the newly appointed members of the Scientific Council immediately pressed the Commissioner about their 'authority over the activities of the agency'; about 'flexibility and lack of "bureaucracy" in the operation of the ERC and in its grants'; about how to adjust 'the structure and legal form' of the ERC, if necessary; and about the proper support of the group with resources.[9] Obviously, since almost all of the legal questions were still negotiated in the political realm of the EU, Potočnik had little concrete to say. As the Scientific Council's politically savvy members certainly reckoned, this opened a unique window of opportunity: as long as the design of the future ERC was not clearly inscribed in the legislation, and as long as its practices had not formed yet, the Scientific Council could still greatly influence key parameters of its operations. Much of its actual power and influence would depend on routines and informal ties yet to be established.

Over the next fifteen months, the first generation of Scientific Council members set up the ERC's core values, its instruments, its mode of operations, and also its rationale. Helga Nowotny later referred to the 'crucial founding years' of the ERC, which had formed an 'institutional memory'.[10] To begin with, the new group had to forge a 'common understanding of aims and strategy' among its members. In its initial session in October 2005, Norbert Kroó, as the most senior member, took over as interim chair, and for the first couple of hours, a messy and confusing discussion unfolded – certainly a result of the fact that people with different levels of understanding and context expressed their thoughts and, inevitably, also exposed their egos. Different ideas buzzed around the room, 'without reaching any firm decisions', as the minutes recall.[11] It came to the point where one prominent member arose and declared that, 'if it goes like this I quit.' Allegedly, this was a wakening call that got the group going, and the atmosphere would increasingly become positive from there on.[12]

In his speech, Commissioner Potočnik had cherished his audience as 'a group of women and men of outstanding competence; you represent an enormous breadth and depth of experience, understanding and imagination.'[13] This was hardly an overstatement. The group consisted of three Nobel Laureates and winners of other important accolades across the board of science and scholarship. Basically, they were all members, and often also leading functionaries, of European or international academic associations in their respective fields; thus well known among their peers and also fully practised in the role of academic gatekeepers. Maybe even more important was the political experience of many of those in the group. Some had been appointees to political positions, such as Michal Kleiber; others had been involved in running important academic institutions at their respective national level, such as the Royal Society in the UK (Bob May) or the Hungarian Academy of Sciences (Norbert Kroó), or in Europe, such as the EMBL (Fotis Kafatos).

Some had been advisers to the European Commission or lobbyists of some sort, with Helga Nowotny chairing EURAB and Jens Rostrup-Nielsen chairing the European Industrial Research Management Association (EIRMA), while several others were members of the grass-roots organization EuroScience. It also brought in some highly distinguished scholars from the humanities, such as Salvadore Settis (who had been director of the Getty Research Institute, Los Angeles, among other positions) and Alain Peyraube (who was in the highest echelons of the French CNRS and official in the French ministry of science). The group was much less well placed in regard to business relations. Only one of its members, Rostrup-Nielsen, had long-time experience working at a non-academic research facility.

What was the most important attitude of the Scientific Council as a collective body was that its members would come to an understanding of their role: to make use of their standing and their networks in order to further establish the ERC, and not to represent their national or disciplinary interests within the body. To meet the goal, the Patten Committee had set up criteria for selecting members of the group: expertise, reputation, broad scientific scope, and also the willingness to participate in their personal capacity and 'independently from any political, governmental, industrial or other interest'.[14] In addition, the committee also maintained that the group 'should be limited to around twenty persons' 'to ensure manageability and [. . . .] to avoid any tendency towards a "juste retour" mentality'.[15] Most of the members of the Scientific Council quickly grasped the role that was expected of them. To act in the interest of science became the core conviction of the Scientific Council as a group; because 'it is only by supporting science on a European basis rather than a country-wide basis that you open your pool wide enough to attract the very best people'.[16]

In early 2007, the Scientific Council wrote a report 'on the occasion of its formal establishment', consciously informing the European Commission that it had been 'working diligently for more than one year in a preparatory effort which will have determining influence on the ERC's subsequent operations'.[17] The sober yet determined language fitted the self-perception of its members; more importantly, the summary also indicated the self-assertive role that the Scientific Council had developed for itself in those months. The notion of the 'founding members' would become a subtle yet palpable distinction among people trained to relate mostly in symbolic ways, keeping apart those involved with the implementation of the ERC and those who came later (and, by implication, had not gone through the same formative experience).[18]

Looking at the service time of all fifty-two members of the Scientific Council thus far (as of late 2015; see also Figure 4.2), the initial group had indeed been longer in office than their successors. But this was due to the fact that the new body was formed in late 2005, though it was formally established only in the beginning of 2007, which is why their four-year terms ran until 2010. Even so, a few initial members dropped out earlier – one because of health reasons

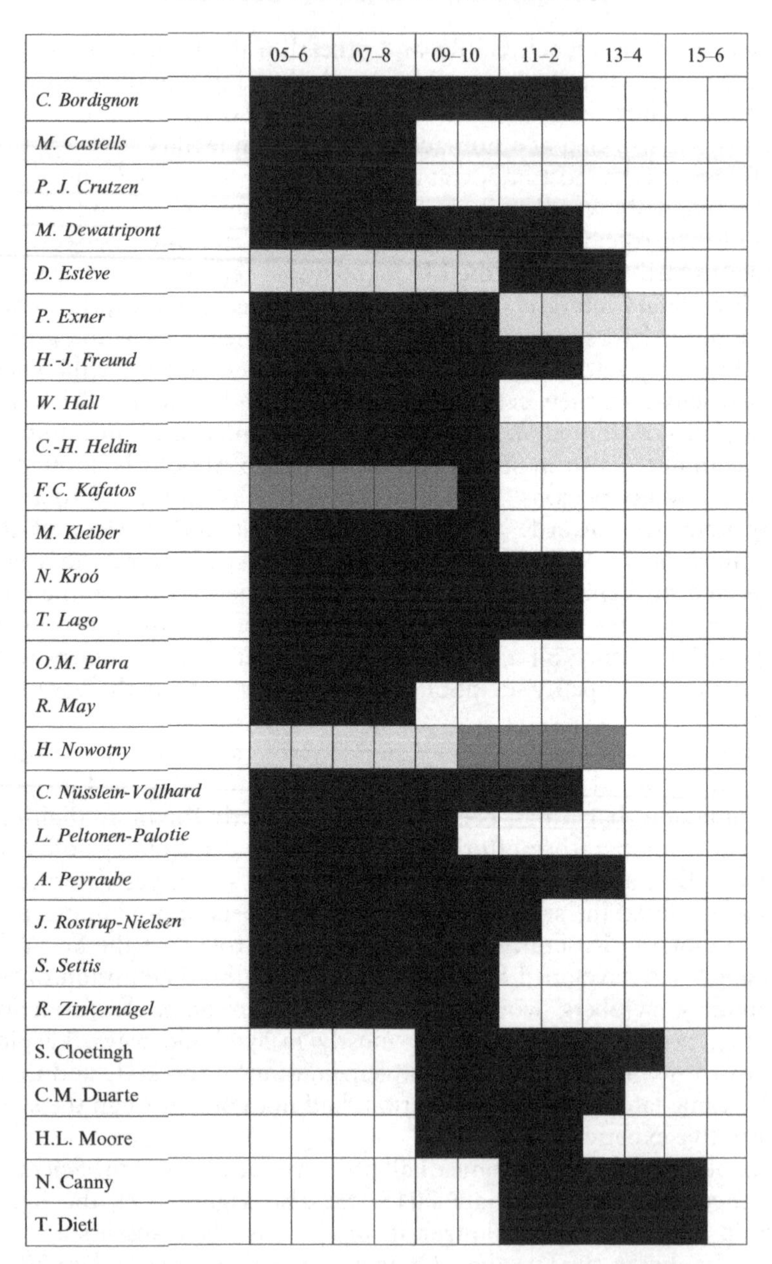

Figure 4.2 Members of the ERC Scientific Council, 2005–2016

Compiled by the author; in italic: initial members of the Scientific Council; black bar: member; dark grey bar: chair; light grey bar: vice chair.

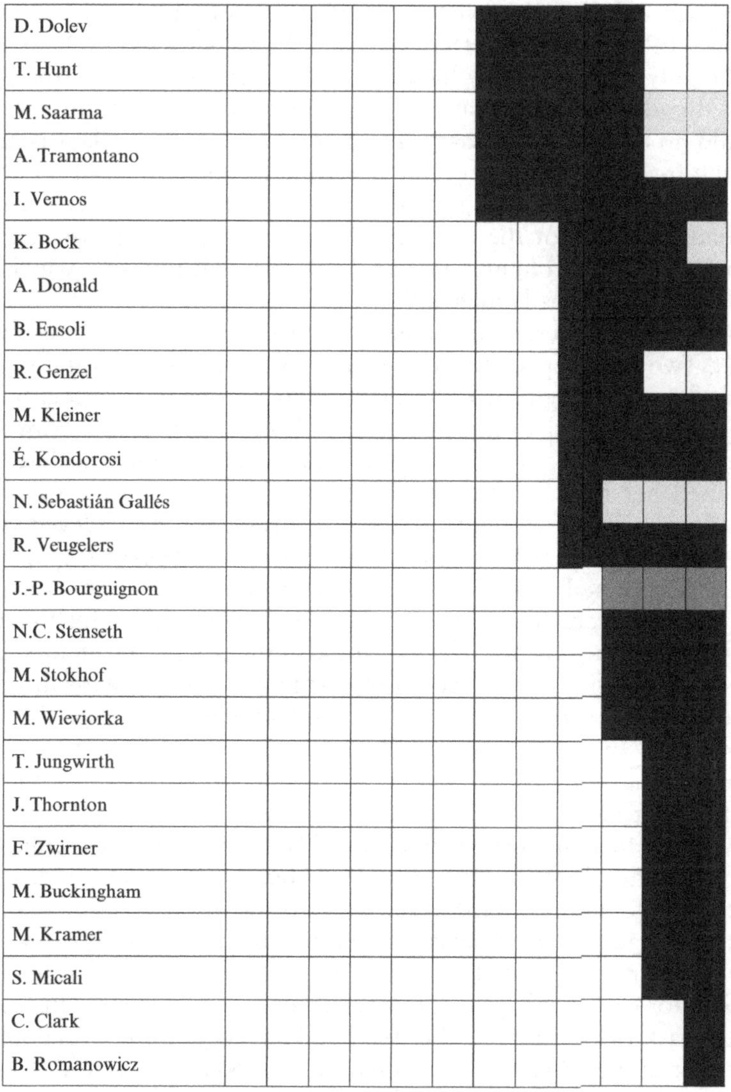

D. Dolev
T. Hunt
M. Saarma
A. Tramontano
I. Vernos
K. Bock
A. Donald
B. Ensoli
R. Genzel
M. Kleiner
É. Kondorosi
N. Sebastián Gallés
R. Veugelers
J.-P. Bourguignon
N.C. Stenseth
M. Stokhof
M. Wieviorka
T. Jungwirth
J. Thornton
F. Zwirner
M. Buckingham
M. Kramer
S. Micali
C. Clark
B. Romanowicz

Figure 4.2 (continued)

(Peltonen-Palotie), but some of them because of failing interest in the project ERC.[19] Paul Crutzen and Bob May, both Nobel Laureates and highly esteemed, may have come to the conclusion that the ERC was already well underway and that lending their name to the cause would no longer be necessary. The case of Manuel Castells was more vexing for his colleagues on the Scientific Council – one of the most prominent scholars on globalization and the knowledge society, he had been one of the intellectual masterminds behind the Lisbon Strategy and thus a figurehead for the broader ideas within which the ERC campaign was launched.[20]

Immediately after the first meeting, the Scientific Council elected its chair. Two members competed for the position: Fotis C. Kafatos, one of the most distinguished life scientists in Europe with special interest in malaria and former director of EMBL, and Helga Nowotny who was equally renowned in the social sciences and who had just declined her membership of EURAB. An electronic vote was cast, based on a simple voting procedure; it turned out even: eleven votes for Kafatos, eleven for Nowotny.[21] Not yet properly set up, the Scientific Council was on the edge of being defunct. The two contenders quickly amended the rules by personally committing for the next round that, whoever was elected would automatically be endorsed by the other candidate who would then stand as deputy. With this sign of unanimity, Kafatos was elected, and Nowotny became the first Vice-Chair. In another round, Daniel Estève, a French physician in quantum mechanics and research director at Saclay Nuclear Research Centre near Paris, became the other deputy.

A few informal rules emerged quickly for the further conduct of the Scientific Council. One stipulated that, in the leadership, three scientific 'domains' – natural sciences (for now: Estève), life sciences (Kafatos), and social sciences and humanities (Nowotny) – would have to be represented. This meant not only an authoritative spokesperson for each respective domain, but also it put someone in charge of the task of putting together the ERC panels in those broad areas. Another informal rule was the division of labour between leadership and ordinary members. The Chair and his/her two deputies would bear the brunt of the implementation work; the others would contribute by providing feedback and crucial links to their respective academic 'tribes' as well as their political networks. The Scientific Council report of early 2007 later stated that the 'leadership' of the Scientific Council had 'devoted a very substantial effort to the organization and conduct of the [Scientific Council's] work'.[22] The election also brought to the forefront the two persons who would have a lasting effect on shaping the ERC. Kafatos and Nowotny were intuitively aware that the numerous legal provisions upon which the ERC would be established would create a compound of entities standing side by side but not necessarily integrated. In order to ensure that the Scientific Council would establish some grip, the latter would need a robust linking between the operative unit in Brussels and the twenty-two scientists voluntarily contributing, and also need to make sure that there would be some safeguarding provisions between the ERC and the parental Commission. As Kafatos wrote in a memo to the Scientific Council early in 2006,

> the success of the ERC will utterly depend on effective coordination of its strategic, supervisory and implementation aspects, leading to an integrated operation.[23]

Kafatos imagined that 'effective coordination' should be based on three pillars: a person entrusted by the Scientific Council 'to ensure the effective operation of the ERC system as a whole'; a culture of 'reciprocal advice' between the European Commission and the Scientific Council when appointing new leadership positions; and the formation of 'two mixed collegial bodies'.[24] The challenge was not so much that the three measures had to be put in practice by the Scientific Council, but that they also had to be accepted by the Commission – which, ultimately, meant that they were to be incorporated in the legal base of the ERC that was, at this point in time, still negotiated by the European institutions.

After its first meeting, the Scientific Council duly embraced 'the establishment of an Executive Agency for the implementation of the ERC' in a public statement (as it was certainly expected, and maybe even prepared by the Commission). But Kafatos and Nowotny did not miss the opportunity also to flex the Scientific Council's muscle by announcing, in the same statement, its intention to recruit a secretary general.[25] That was a shrewd move: the Commission dearly needed the Scientific Council's public support at the time – as shown in Chapter 3, the major motivation to convene the group at this early stage was to make a convincing argument for the advantages of its preferred implementation model. By literally chaining its endorsement to a position that was not foreseen in the Executive Agency blueprint, the Scientific Council could hope to gain a foothold in the otherwise secretive and inescapable machinations that would be set in motion by the Commission apparatus once the next edition of the Framework Programme was formally adopted.

In order to follow through and present its proconsul as a *fait accompli*, the Scientific Council was in a hurry: the position had to be filled before the next edition of the Framework Programme started – that is, before the turn of the year. At the same time, the Scientific Council would have to identify someone appropriate for this job – that is, someone of impeccable status and ready to move to Brussels. Actually, Fotis Kafatos already had someone in mind. In the autumn of 2005 – before the first meeting of the Scientific Council, but then already formally appointed by Potočnik – he and two of his colleagues had written a memo to Commissioner Potočnik, taking 'the liberty to suggest for your consideration Professor Ernst Winnacker, as a potential director of the Executive Agency of the ERC'.[26]

And yet, to avoid any unfavourable impression, the Scientific Council decided to set up a fair and transparent competition, including an open call for applications. It also imposed a tight schedule, with the plan to identify a shortlist of suitable persons in the summer and, after interviewing them, to cast an electronic vote.[27] Despite this procedural junction, Fotis Kafatos could be confident of getting his candidate through. After all, Winnacker had done impressive work at the DFG; he had been a fervent, though somewhat hapless, ERC advocate from the very beginning; and, over the past years, he had established a unique European-wide funding stream at the ESF called EURYI, which served as a template for one of the funding schemes that the Scientific Council was about to develop for the ERC. No other contender could offer a similarly impressive list of accomplishments.

Yet the stipulated procedure had to be followed through. The secretary general position was publicly announced in late April 2006 and the call yielded some twenty applications. A designated recruitment committee from within the Scientific Council members came up with a shortlist of four names. The interviews took place on 20 July in London, at Imperial College, Kafatos' host institution. The Chair and his team had meticulously prepared the event; however, Winnacker made a fairly miserable impression during his presentation,[28] and much to the frustration of Kafatos, 'no clear consensus about the candidates was formed'.[29] In vain, Kafatos agreed to meet with each short-listed candidate for a personal in-depth interview, after which he again urged the others in the Scientific Council to stand behind Winnacker.[30] The result of the electronic vote did not support the chair: Andreu Mas-Colell, a Spanish economist with a brilliant academic career in the US and ample political experience, received eleven votes, and Winnacker only nine (one vote was given to another candidate, and one vote was spoiled).[31]

The situation was difficult for Kafatos: in order to leverage a 'clear-cut victory', he had *de facto* committed his own chairmanship to Winnacker. Now, in the event of a second round of voting, it seemed even more likely that Mas-Colell would receive the majority of votes, rather than Winnacker bouncing back. Kafatos and Nowotny risked a change of tactic. Warning that a second round of voting 'might also end up inconclusively', they argued for splitting the five-year term of the Secretary General: Winnacker should take the first two years, with Mas-Colell stepping in afterwards. 'Of course such split terms are not unprecedented at other institutions', the memo reassured the other Scientific Council members, and proposed that the idea should be implemented 'as soon as it has been endorsed by twelve members of the Scientific Council, including ourselves.'[32] Upon such determinacy by its leadership, the majority acquiesced, and Winnacker and Mas-Colell both agreed to the solution.

In the press statement announcing the appointment of Winnacker and Mas-Colell, Kafatos praised the 'broad Scientific Council consensus', while Nowotny referred to 'its farsighted view and balanced assessment of what the ERC will need in the near future.'[33] That was a spin for the broader public as well as for the European Commission, which had been excluded from the procedure, and remained, for the moment, rather indifferent. However, behind the façade of unanimity, the Scientific Council had proven to be at times difficult and refusing to be steered by its leadership; and, more seriously, the compromise would leave Winnacker somewhat doubtful about his role.[34]

4.2 Internal policies

In the beginning, the Scientific Council's only regular point of contact with the European Commission services was a daring official. In the early 2000s, William Cannell had been among those inside the Commission becoming increasingly critical of the FP format. In his opinion, most of its programmes suffered from two weaknesses: for one, scientists perceived them as overly bureaucratic because of the many forms they were forced to fill out; and second, the funding calls were too narrow, not giving researchers enough freedom to come up with their own ideas.[35] When he was tasked to establish the 'New and Emerging Science and Technology' (NEST) programme for the sixth edition of the Framework Programme,[36] Cannell made two crucial adaptations to how things were normally done: he brought the financial oversight and the programme officers together in one administrative unit. This way, the accountants, who were responsible for supervising the financial aspects of FP-funded projects, could be made familiar with the peculiarities of scientific research, and as a result, scientists conducting a project funded by NEST were released from much red tape. The other innovation was that Cannell introduced three main funding streams, thereby positioning NEST as a mostly bottom-up programme. As its first Work Programme stated:

> [. . .] if Europe is to maintain a truly dynamic research capability, it needs not only to support the critical research areas for tomorrow, but also to seek out the most promising opportunities for the day after.[37]

An inspiring administrator, Cannell also had a knack for identifying like-minded people within the Commission services; his unit thus became a hotbed for innovative spirits and acquired deep knowledge of how to make best use of the legal provisions of the FP format for organizing funding in a researcher-friendly mode. NEST became an exemplary test case, and it gained a lot of credibility among those in the scientific community who had been exposed to the programme. Its layout had many overlaps with what the ERC advocates perceived as the main principles of the future ERC. There was one weakness, however: with only some € 215 million allocated between 2002 and 2006, it was far too small to cover all fields of science, and to gain broad acclaim.

The ERC offered an entirely different perspective, and, after some initial reluctance, Cannell and his team accepted that the brand name NEST would have to be sacrificed. Cannell's unit, or the 'Secretariat', as it was soon dubbed, turned out to be an ideal match to the Scientific Council: based on their experience with the NEST programme, this unit had the sensitivity required for dealing with the staunch and sometimes stubborn ERC advocates. It was no surprise, then, that, in the collective memory of the first generation of Scientific Council members, the first year remained that of a happy period: the collaboration with Cannell was mostly a fruitful one, and it left the Scientific Council on its own when setting up the operational details of the ERC.

Yet there were also irritations; the most significant being the European Institute of Innovation and Technology (EIT). The EIT was, like the ERC, an attempt to grapple with the failing prospect of getting the implications of the Lisbon Strategy going.[38] In many respects, however, it was the opposite to the ERC: it 'emerged on the European agenda in response to Manuel Barroso's political will to set his footprint as the new Commission President'.[39] Given that it was the sole initiative of a politician, the EIT idea found little sympathy among stakeholders; yet it was implemented based on exactly the same legal provision ('article 171') that the European Commission had, only some months ago, deemed to be inappropriate for the ERC. The Scientific Council, for its part, made great efforts to define the ERC's relationship to this new body; it provided some feedback on the EIT but otherwise maintained that the 'distinction between the missions of the ERC and EIT must be carefully preserved and effectively presented to stakeholders', as an internal position paper stated.[40]

Positioning the ERC to other (new) players in the theatre of European research policy was one thing; the other was to shape key parameters of the ERC future operations. With the two consecutive Secretary Generals appointed and publicly announced, Kafatos could establish another pillar of his coordination policy by setting up a format of regular meetings in order 'to maintain the continuity of operational aspects'.[41] Convening in Brussels, the 'ERC Board' (as it would be called) would bring together the Commission management with Kafatos, his deputies, and the Secretary General.[42] It met for the first time in late November 2006 and adopted 'its operating methods', of which the most notable was that 'the agenda of the meeting is established under the authority of the President'.[43]

Almost a year after their initial gathering, the Scientific Council met for its sixth plenary meeting. The session, taking place in London, started with an 'informal session on recent developments',[44] in which Fotis Kafatos informed his colleagues of news he himself had received only a few days ago: the Commission was about to reshuffle its respective Directorate General handling R&D in order to be prepared for the administrative challenges of the next edition of the Framework Programme to be enacted from January 2007 on; for the ERC specifically, a new Directorate (carrying the letter 'S') within the DG would be created, 'which will be transformed into the Executive Agency (EA)'. Subsequently, the Director General of DG Research arrived in person and introduced Jack Metthey as head of the new administrative unit.[45]Although the notes of the meeting do not record discussion during the internal session, nor a qualifying remark on the exchange with the Director General, the situation must have been awkward. Up until then, the Scientific Council's almost exclusive line of communication had been via William Cannell. Among others, this concerned updates on the progress of the 'political and financial developments' of the upcoming edition of the Framework Programme.[46] While the grounds on which the ERC was supposed to operate were becoming reliable, the Scientific Council was working at arm's length from the intricate political negotiations – a convenient situation, which, as an important side effect, also formed strong ties and allegiances among its members.

Now, this happy state of affairs had come to a rather abrupt end. The Scientific Council was confronted with an important structural overhaul, which would directly affect its further work, and with a new main liaison officer. Jack Metthey was an unknown.[47] It didn't help that the Commission had ignored Kafatos' idea of 'reciprocal advice' (he had obviously wished for another candidate – William Cannell – to head the new directorate). The 'Secretariat' remained the Scientific Council's first point of contact and continued to deal 'with questions of overall strategy and communications for the ERC';[48] but it was now only one out of the four units of the new directorate, and Cannell's own position was pushed back.

By then, the legal basis of the ERC compound was becoming clear: it would consist of several texts, following a hierarchy of four layers,[49] depending on different degrees of specification, but also on the authority formally adopting it (see Figure 4.3). On top was legislation that was the basis upon which the ERC could be set up. Next was the 'first tier of specific legislation' with the 'Ideas Specific Programme' and the 'rules for participation', the latter detailing 'the minimum conditions' under which 'undertakings, research centres and universities' could apply for funding.[50] As is common in the European polity, the texts at the top of the hierarchy as well as in the first tier below were adopted by the EU legislative powers. In order to implement the respective policy (in this case: the 'Ideas Specific Programme'), the European Commission was expected to install the 'second tier legislation'.

Typically, the provisions in the first tier legislation for a specific instrument like the ERC would remain very general. That would leave much room for interpretation and leeway for refining the legal obligations for implementing the ERC compound in the second tier. Thus, when the Commission started drafting the second tier texts in the second half of 2006 (it started doing so in parallel, as drafts of the legal base and the first tier were already under discussion), Kafatos' policy of coordination was at stake. While the 'Rules for Submission' detailed how the ERC would conduct its business of allocating funds to research, the 'ERC Decision' made crucial provisions about the Scientific Council, its setup and renewal; a separate Commission Decision on setting up the ERC Executive Agency complemented the tier (it would only be adopted a year later). The Commission informed

Title	Formal notation	EURLEX	Adopted
Legal basis			
Executive Agency Regulation	Council Regulation laying down the Statute for executive agencies to be entrusted with certain tasks in the management of Community programmes	2003/58	19.12.02
FP7	Decision of the European Parliament and of the Council concerning the Seventh Framework Programme of the European Community for research, technological development and demonstration activities (2007–2013)	2006/1982/ EC	18.12.06
First tier legislation (legal provision)			
Ideas Programme	Council Decision concerning the specific programme: 'Ideas' implementing the Seventh Framework Programme of the European Community for research, technological development and demonstration activities (2007 to 2013)	2006/972/ EC	19.12.06
Rules for Participation	Regulation (EC) of the European Parliament and of the Council laying down the rules for the participation of undertakings, research centres and universities in actions under the Seventh Framework Programme and for the dissemination of research results (2007–2013)	2006/1906	18.12.06
Second tier legislation (legal implementation)			
ERC Decision	Commission Decision establishing the European Research Council	C2007/134/ EC	2.2.07
	Commission Decision amending Decision 2007/134/EC establishing the European Research Council	C2011/12/ EU	12.1.11

Figure 4.3 Four layers of legal provisions of the ERC

Compiled by the author.

Title	Formal notation	EURLEX	Adopted
ERCEA Decision	Commission Decision setting up the 'European Research Council Executive Agency' for the management of the specific Community programme 'Ideas' in the field of frontier research in application of Council Regulation (EC) No 58/2003	C2008/37/EC	14.12.07
Rules for Submission	ERC Rules for the submission of proposals and the related evaluation, selection, and award procedures for indirect actions under the Ideas Specific Programme of the Seventh Framework Programme (2007–2013)	C2007/2286	6.6.07
	Commission Decision amending Decision C(2007) 2286 on the adoption of ERC Rules for the submission of proposals and the related evaluation, selection and award procedures for indirect actions under the Ideas Specific Programme of the Seventh Framework Programme (2007 to 2013)	C2010/767/EU	11.12.10
Work Programme	ERC WP 2007	C2006/561	26.2.07
	ERC WP 2008	C2007/5746	29.11.07
	ERC WP 2009	C2008/3673	23.7.08
	ERC WP 2010	C2009/5928	29.7.09
	ERC WP 2011	C2010/4898	19.7.10
	ERC WP 2012	C2011/4961	19.7.11
	ERC WP 2013	C2012/4562	9.7.12

Figure 4.3 (continued)

Title	Formal notation	EURLEX	Adopted
Delegation Act		n.a.	
Model Grant Agreement		n.a.	
Execution and guidance texts			
ERC Guide for Applicants		n.a.	
ERC Scientific Council Rules of Procedure		n.a.	
ERCEA Annual Activity Report		n.a.	
Guide for Peer Reviewers		n.a.	

Figure 4.3 (continued)

the Scientific Council about the status of drafting the various texts at its plenary meetings, and asked for comments.[51] Yet for a group of voluntarily assembled scientists, it must have been difficult, if not impossible, to supervise meticulously the various drafts in order to ensure that its own interests were honoured.

Kafatos and his colleagues had set important impulses for shaping the ERC's practical governance, yet they failed in anchoring the Board in the second tier legislation, as it was not incorporated in the 'ERC Decision', which otherwise explained in more detail the remit of the Scientific Council. What would soon become the most important interface for making the different entities of the ERC operating together remained informal for the next few years. The 'ERC Decision' explicitly mentioned the secretary general position. However, even after the draft had been discussed during the Scientific Council's meeting in London and the group had asked for changes 'as regards the term of membership and the term of appointment of the Secretary General',[52] the final version remained vague about its remit: it gave assurances that he (or she) would be independently selected and should also ensure 'effective liaison with the Commission' and the agency; it also held that his or her tasks should 'be defined by the Scientific Council'; but it made no provisions for the position's role in the overall composition of the ERC compound.[53]

The results of the negotiations of the second tier legislation were mixed. It acknowledged some extraordinary positions, and it made some concessions of important symbolic value – most notably, the chair of the Scientific Council would, from now on, also be the President of the ERC.[54] But it did not provide real bridges. And certainly, it did not assert the Scientific Council's wish to be in command. The reshuffling, the few concessions in the second tier legislation, and the limited efforts to take up Kafatos' three pillars of coordination policy were certainly due to the fact that the Commission was under time pressure and was entering new territory. But it was also a sign that the Commission would follow a strict and formalistic approach for setting up the ERC's working procedures, and would regard the 'two key structural components of the ERC – an independent Scientific Council and a dedicated implementation structure' as mandatory.[55]

For a bureaucratic body prone to professionalism and objectivism and with its (sometimes overstated) disgust for political preferences, such strictness may have been self-evident. Still, after the honeymoon period during the early meetings of the Scientific Council, the Commission's brusque way of informing the Scientific Council in London must have come as a surprise. The impression that the Commission's attitude towards the ERC was cooling was not completely off the mark. In the second half of 2006, the ERC had changed from a political project to an administrative issue. While the former had demanded strategic vision and tactical skills in different sets of negotiations, the latter concerned the implementation of an alien (or, at least, new) element into the research funding machinery at the Directorate General already working at full stretch.

The change was further enforced by the fact that this machinery was now under political scrutiny. With the seventh edition of the Framework Programme, the Commission had been tasked to distribute a substantially increased budget – a great political achievement, but also a huge administrative burden. More financial means implied more contracts and grants with an increasing number of beneficiaries, i.e. universities, research units, non-, and for-profit, enterprises spread across a recently enlarged Union, many of which had only limited experience with the Commission's accounting standards.[56] Not surprisingly, then, the Commission perceived its research funding policy at risk of being exposed to political allegations of wasteful spending, and made attempts to simplify the rules for participation, but also to tighten its control measures.[57]

In the informed research policy community, the changes were connected with a prominent and, at least symbolically, rather infelicitous change in leadership: in late 2005, Achilleas Mitsos had been rotated to another post and was succeeded by José Manuel Silva Rodríguez, or Cuqui, as he was nicknamed around Brussels. Like Mitsos, Silva Rodríguez belonged to the small circle of 'powerful political bureaucrats seeking to further the intents of their organization';[58] yet partly because of his heritage, he would soon be blamed by members of the Scientific Council, as well as by some Commission people, that he had unnecessarily complicated the ERC business. Silva Rodríguez had built his career in the DG Agriculture,[59] certainly one of the toughest directorates in the European Commission to administer. His credentials cannot be in doubt: while Mitsos had guided the ambitious project of setting up the seventh edition of the Framework Programme (implementing, among other innovations, the ERC), he had accomplished a major reform of the Common Agricultural Policy.[60]

There were two circumstantial aspects that necessarily impacted the way the new Director General perceived the ERC. One was that, unlike Mitsos who had made the ERC a special item on his political agenda, Silva Rodríguez had not been much concerned with research policy thus far; consequently, he must have regarded the ERC more indifferently than his predecessor. The other aspect was that, when Silva Rodríguez started at the DG for Research, programming of the Framework Programme's new edition was already broadly agreed upon, and he would have calculated that his term would probably not last long enough to initiate the next one. With his capacity for policy-making effectively restricted, he focused on a rather bureaucratic, but nonetheless highly aspirational goal: to bring down the so-called 'error rate' in the accuracy of cost claims of funding contracts. For a prosaic manager with no emotional attachments to a specific project, that included all policy instruments under his control – also the ERC.

Silva Rodriguez' initial actions concentrated on reshuffling the Directorate General and on management appointments. Picking Jack Metthey showed a good understanding of what was needed for the intricate position of a liaison officer with the Scientific Council. Metthey easily won the group's trust, including that of Winnacker, while ensuring the transition of the ERC compound into operations. Similarly, establishing an entire Directorate as a 'dedicated implementation structure' (DIS), which would then have to be transferred into an Executive Agency, was a step towards giving the ERC the administrative support that it deserved. Yet the Director General also appointed people from within the Commission services to the new directorate who did not share the optimistic vision of the ERC as being a 'revolutionary development' for European research policy. Their ambition was to align the ERC along the same regulatory standards that had been developed for the rest of the Commission's research funding lines. As the Commission leadership perceived it, those 'watchdogs' would contribute to the ERC's operational and financial compliance; but, because they told the Scientific Council what it was not allowed to do, they conveyed a rather idiosyncratic image of the Commission.[61]

— 5 —

STATE OF CRISIS

The ERC was officially launched in a celebratory event in Berlin in February 2007, during the German EU Presidency. The newest edition of the Framework Programme was formally adopted, and the ERC had just announced its first funding call. During its stay in Berlin, the Scientific Council had an audience with the German President and met other luminaries. At the main event, Chancellor Angela Merkel gave an opening speech, wishing the ERC's Scientific Council 'every success, good luck, stamina and steady nerves – and also a fair share of fun.'[1] For now, there was love, peace, and harmony. Only Angelika Niebler, who had been the Rapporteur to the European Parliament on the ERC, referred to 'the challenges to be taken up',[2] a reminder that the current institutional solution for the ERC was without good precedent and still contested.

An internal report of the launch event summarized that, in the eyes of the informed public attending the opening celebrations, the ERC 'appeared to be an already accepted new research actor'.[3] Unlike what its advocates had hoped for, however, the ERC was not a single entity; rather, it was built on several tiers of legal texts, starting with the seventh edition of the Framework Programme and further defined in the 'Ideas Specific Programme', effectively creating three only loosely connected entities. On the legal drawing board, it may have seemed as if those entities would genuinely complement each other. However, the result was that the ERC rested on a mixture of legal provisions that, in comparison to the NSF for example, were more restrictive and highly ambiguous at the same time (see also Figure 5.1).

They were more restrictive not only because the ERC was formally running only for the time of one edition of the Framework

	NSF[4]	ERC[5]		
Legal basis	Single public act	Several tiers of decisions[6]		
Term	Indefinite	For the time of seventh edition of FP		
Structure(s)	'Independent non-regulatory commission'	'Ideas Specific Programme'	'Scientific Council'	'Executive agency'
Reporting	US President	Member states	European Commission	DG Research
Leadership units	National Science Board, Director	Programme Committee	Chair; Secretary General	Steering Committee, Director
Delegation Principle	Fiduciary	-	Fiduciary	Agency

Figure 5.1 Comparison of NSF and ERC governance, 2010

Compiled by the author.

Programme, but, more significantly, it was subjugated to the regulations both of the FP format and of the 'Executive Agency' model. And it remained ambiguous, because the three entities were rather hastily glued together, which meant that each came up with its own leadership units, creating a hardly comprehensible structure with different obligations of accountability: while the Programme Committee of the 'Ideas Specific Programme' would remain a pro forma body,[7] the management of the executive agency was strictly based within the Commission and reported to the Directorate within the DG Research to which it appertained; the Scientific Council, finally (and somewhat surprisingly), was being held accountable to the Commission – without any further obligations.

There have been different attempts to characterize the ERC: as a 'hybrid' between programme and institution,[8] as a 'dual construction',[9] and so on; those codes rather express the perplexity of scholars in the face of the wobbly construction of the ERC compound. For now, it is sufficient to understand that the Scientific Council and the Executive Agency – as the two entities implementing the paramount Specific Programme – rested on very different principles of delegation. The 'lean and cost-effective dedicated implementation structure'[10] was based on a legal provision that followed the well-known principal-agency model, which is used 'to reduce decision-making costs'. The Scientific Council, on the other hand, was created on the principle of a fiduciary, namely 'to enhance the credibility of policy commitments'.[11] To complicate matters, the Scientific Council remained a fiduciary without its own resources, while the 'dedicated implementation structure' (first, Directorate S, later the ERC Executive Agency) was an agency with two contrary principals, one being (legally provided) the Commission, the other being (politically imposed) the Scientific Council.

It is hardly surprising, then, that the tensions and frictions that inevitably occurred in the beginning of the ERC within this compound were hardened by different expectations and imaginations that those two delegation principles created; for a while, it was an open question whether they were merely start-up problems or whether they would be more persistent. Occasionally, the Scientific Council leadership took its frustration to the public, mostly in order to put pressure on the powerful European Commission to loosen its 'overly strict control culture'.[12] In the end, and contrary to what many advocates had anticipated, the ERC compound was not failing. The reason for this was, primarily, that the Scientific Council had 'forged a trusted relationship with its Executive Agency and a highly professional staff' – but not '[a]fter overcoming many initial difficulties',[13] at which this chapter will take a closer look.

5.1 Deep Commission

The first year of its establishment had developed a division of labour between the Scientific Council and the administrative staff, first in Cannell's unit (the 'Secretariat') and then in Directorate S: the Scientific Council members, with the ambition to craft 'a clear and compelling vision' for the ERC,[14] stipulated rules by looking at procedures from the perspective of their expected impact and not from the angle of how they fit into the existing legal provisions. And the administrative branch under William Cannell and later Jack Metthey implemented those rules, because its staff knew the intricacies of the financial and the Framework Programme regulations and found a way to cut through them.

This *modus operandum* remained in place after the ERC began its operations in earnest in 2007, but priorities were shifting. The Scientific Council's focus was no longer on implementation but on monitoring; the observed problems and obstacles were reported to the administration; Metthey and his management team readily agreed to correct them, and sometimes, this worked efficiently in the eyes of both entities. For example, when the idea of bringing applicants for ERC funding to Brussels in order to conduct interviews was proposed, the initial response from the administration was that 'there is no legal base for reimbursing the travel costs'. The obstacle was brought to the highest echelon of the Commission, but the solution was found thanks to diligent and creative work by 'Directorate S'.[15]

However, because the administrative branch often had to secure legal (and sometimes also political) backing for changes, solutions often took time; the Scientific Council, which was not involved in the consultations with other Commission services, had no good understanding of what caused red tape and delays in implementing changes. Problems usually could be resolved, even though not as quickly as the Scientific Council had hoped.[16] Enduring frictions were not concerned with procedural issues within the ERC decision-making procedures, but with internal governance. Mostly, this revolved around three topics: the position of the secretary general; the Scientific Council's privileges; and its leadership's endowment.

After the Scientific Council had appointed Ernst-Ludwig Winnacker and Andreu Mas-Colell, consecutively, the secretary general position was formally included in the legal text with the task to 'assist the Scientific Council in ensuring its effective liaison with the [agency] and with the Commission, [and] in monitoring the effective implementation of its strategy and positions.'[17] Winnacker, full of 'enthusiasm for the ERC', moved to Brussels in early January 2007, eager to get started.[18] Upon entering uncharted territory, he may have expected that, as a replacement for the clear executive function, and also quite a few amenities that he had enjoyed as President of the DFG, his new position at the ERC would come with far-reaching responsibilities, including the nascent organization's future strategic scope. But Winnacker quickly came to realize that, just as the ERC was not an integrated entity, the Secretary General had 'no formal administrative functions or responsibilities' at all.[19] To make things worse, the European Commission services had not properly prepared for his arrival, and continued to behave strictly in a non-committal way towards him.

Despite these difficulties, Winnacker contributed greatly to the building up of the ERC's operations. A distinguished scientist and experienced manager of research-funding organizations (not only the DFG, but also the EURYI programme), he knew by heart the intentions of the policies that his colleagues in the Scientific Council had devised, and how to translate them into working. Quickly, he emerged as guiding figure in the conduct of the initial ERC funding calls, instructing and supervising those of the newly recruited staff who were dealing with the administrative side of the decision-making procedure (carrying out eligibility checks and assignment of proposals, dealing with review panels and remote reviewers, and the carefully drafted policies around those, and related, issues). Many of the 'Scientific Officers' at ERC premises had held university positions before; they were close to the scientific community, probably had known Winnacker by name and appreciated his experience in executing firm peer-review standards as well as his professional collegiality.

This admiration was not shared in other parts of the Directorates.[20] Even after November 2007, when eventually his job was formally settled and his first salary paid, Winnacker's hands remained tied: he was not allowed to participate in management meetings – instead, Metthey established a separate weekly gathering with him and the Directorate's heads of unit.[21] Throughout his term, Winnacker was obliged to report his absences and had to be formally signed off by Kafatos. It is not clear if this was on purpose or just due to bureaucratic thinking, but the way the Commission handled the Scientific Council's proconsul must have been perceived as a continued insult; and it was nourishing fears that the Scientific Council was, despite all nice words, tolerated rather than valued.

Another sort of insult was endured by the Scientific Council concerning its privileges in assessing the results of the ERC's operations. Shortly after the summer break 2007 and when the first funding Grant call was just concluded, several members 'complained about leaks of information'. During occasional chitchatting with their peers, they had been confronted with results of the funding call 'that they themselves did not have.' As it turned out, the regular meeting of the Programme Committee of the 'Ideas Specific Programme' – which, as mentioned before, had mainly pro forma responsibilities – had incidentally been scheduled before the Scientific Council plenary. In preparation, the Commission had dutifully reported information on the results to the members of the committee: formally, as Jack Metthey then argued in the unfolding discussion with the Scientific Council, the Programme Committee 'has the prerogative of receiving information first', since its members – as representatives of the member states – were the 'sponsors' of the ERC programme.[22]

The results of the first ERC funding call were expected with great anxiety in many parts of the European research landscape, including ministries and funding agencies. Since the Scientific Council had decided to give full authority on the funding decision to the panels, any reviewer could leave Brussels and spread the results, and it should not have come as a surprise that there were leaks. Yet, having carefully set up the scheme and being thrilled by how the first call was going, the Scientific Council members now felt embarrassed to be confronted with details of the outcome by people who were essentially bystanders. The Scientific Council maintained that it was its prerogative 'to receive any relevant information on time, and before publication.'[23] The concern about its privileges on data access would randomly flare up again on different occasions.[24] On each of those occasions, the European Commission's reluctance to inform would be a piercing reminder to the Scientific Council members about their limited remits.

Finally, there was the issue of substantial support to the Scientific Council leadership. The Commission had introduced a provision in the legal text that 'Scientific Council members shall not be remunerated for the tasks they perform'[25] – the noble assumption was that they would work on an entirely voluntary basis. However, the workload of Kafatos and his colleagues clearly passed any level of what could be done without support. To provide for the workforce, William Cannell had initially come up with a support contract under the NEST programme, which provided the necessary funds to compensate (part-time) the salary of their secretariats at their host institution; in addition, the chair also employed an adviser who helped him with drafting presentations and speeches, preparing policy papers and guidelines, and coordinating forthcoming Scientific Council plenary meetings with the Secretariat in Brussels.

Given the impromptu assignment with which the Scientific Council had come to life and the speed with which it accomplished its foremost obligation, Kafatos may have expected that the Commission would exercise leniency with costs that – from his perspective – were fully justified for the best of the ERC, even if they were not complying with existing regulations; and once the ERC was established, he also may have hoped for a long-term contract under which money for administrative support was at the President's disposal as he saw fit. The Commission leadership readily acknowledged 'the efforts that have been made by the Scientific Council towards the preparation of the ERC', and promised to find 'a long term stable solution in line with the given legal framework.[26] However, things were not so easy to settle, because leniency was not a principle of the Commission's strict set of regulations; and the compromise about how to settle this issue in the future made nobody who was involved happy.[27]

It was Jack Metthey's task to give an the explanation to a reluctant Kafatos, Nowotny, and Estève that they could not expect a 'general subsidy'.[28] Instead, the support had to be adopted in the annual ERC

Work Programme, with the President having formally to apply to it, to 'negotiate' the contract and, at the end of the one-year period, to write a final report with 'deliverables'. The process repeatedly demanded time and efforts that the Scientific Council leadership deemed wasted.[29] But for Kafatos (who greatly relied on his advisers), and for Nowotny (who, as a retired scientist, was not entitled to a secretariat otherwise), this support line was of the utmost importance, as their contributions to building up the ERC would have been radically constrained otherwise. The mixed messages sent by different parts of the Commission – with the DG Research leadership assuring in tone, and the financial officers regularly doubting certain cost claims – were thus received with increasing doubt about the overall sincerity of the Commission. Not to mention that exposing the weakness of their monetary situation hurt the pride of the Scientific Council members and provocatively endangered their efforts.[30]

The three areas of friction remained constant reminders of the awkward position of the Scientific Council within the ERC compound. For the President personally, it was even worse – 'feels exploited', as he noted later; and also, the Commission 'does not realize how deeply offensive it is to have to spend too much time finding solutions to [Commission's] own convoluted rules and mechanisms'. Kafatos was not blaming Metthey and those close to the ERC, but – what he called – the 'deep Commission', meaning the intransigent rules but also 'high-level officials [who] are arrogant, over-confident and oblivious of the hurdles they put the ScC through, in flagrant disrespect for the time and effort of the ScC.'[31]

5.2 We are not there yet

Even more of an impediment to the ERC, in the eyes of the Scientific Council leadership, would be the transition of Directorate S into a separate executive agency. The difficulty started with Commissioner Potočnik struggling to acquire political consensus for setting up the new agency; for a while, he even 'worried about the difficulties in setting up the agencies that this Commission has already agreed in principle to set up'.[32] For Kafatos and his colleagues, all this must have been bewildering. Had not the Commission made a strong argument in favour of the 'Executive Agency' model? And had not the political institutions, with some resistance by certain member states, just agreed to establish the ERC along this path? Why, then, was there suddenly this obstacle?

In Europe's complex fabric of political decision-making mechanisms, one choice was not necessarily synchronized with consecutive steps; and the establishment of organizational bodies under the control of the European Commission fell into the remit of policy-makers who were not really involved with research policy – the College of Commissioners, the Economic and Financial Affairs Council (ECOFIN), and the Parliament's Committee on Budgetary Control. The legal provision for executive agencies had been set up with the goal 'to delegate some of the tasks relating to the management of Community programmes', thereby 'achieving the goals of such [. . .] programmes more effectively'.[33] Yet, as an independent analysis around the time showed, Commission services in general were rather reluctant to delegate tasks, unless 'staff shortages (in number and specialization)' were forcing them to do so. In the case of the ERC, however, 'no staff are planned to be freed, as the activities delegated are totally new.'[34]

Finally, the 'ERCEA Decision' formally established the Executive Agency shortly before the end of 2007.[35] But soon Winnacker and his President detected 'problems and inaction on important issues' that they claimed would threaten to 'seriously damage the ERC'. The reason was that internal Commission guidelines for establishing an executive agency demanded a sequence of consecutive procedures and intricate negotiations within different Commission services: for example, the protocol mandated a director to be in place before any other step (such as hiring staff) could be started. Maybe even more disturbing for Kafatos and his colleagues, they were informed that they would not be involved in several key decisions – most notably, the appointment of the director of the agency. The post was announced shortly after the agency was formally established, but Kafatos and his colleagues then recognized

> that none of the talented people with whom we have worked to date (Robert-Jan [Smits], Jack [Metthey], William [Cannell]) were eligible, as the standard process specifies very precisely the status [. . .] and seniority [. . .] of an EA Director, which none of the three meets.[36]

Over the next months, as meetings with Silva Rodríguez and Potočnik as well as with Commission President Barroso[37] would not produce the results that Kafatos had hoped for, the optimism within the Scientific Council about quickly setting up the agency vanished,[38] and was replaced by frustration, anxiety, and resentment; its chair, in the meantime, grew restless. 'We cannot continue for very long like this', his preparatory memo for another meeting with

Commissioner Potočnik stated, twice.[39] A couple of months later, when invited to speak in front of the national science ministers in the Competitive Council, Kafatos reported 'administrative difficulties [. . .] which threaten the existence of the ERC', due to the fact that 'the Commission has not delivered so far.'[40]

In some sense, Kafatos' efforts to voice his concern were successful. In summer 2008, the agency's Steering Committee was already meeting informally (to 'avoid a cold start of the Agency'[41]), and, since the Scientific Council had demanded that 'one of its members', together with 'a person with long experience in research management as second outside member',[42] should become members of the new body, the Commission had assigned Scientific Council member Mathias Dewatripont and Catherine Cesarsky, renowned astronomer and chair of one of the ERC evaluation panels for the 2007 funding call. Also, Jack Metthey was appointed director ad interim of the executive agency (following a short period of Silva Rodríguez acting as director after the Barroso-meeting, which was not a suitable solution). All three – informal meetings of the Steering Committee, bringing in people from outside the Commission, and appointing a director on an interim basis – were deviations from the Commission standard protocol, and would not have happened without Kafatos' insistence.

Those successes could also be interpreted as a willingness to compromise on behalf of the Commission. As Silva Rodríguez and Potočnik probably saw it, the Scientific Council should have put more trust in the Commission services. Despite some hiccups, the implementation of the ERCEA was deemed to be on track. From the perspective of a professional administrative body, this may have been accurate; to the Scientific Council, however, it appeared as an endless series of alternating delays, complications, and revelations. Kafatos and his colleagues lacked an understanding of the internal procedures and regulations, and their sense of urgency was probably reinforced with Winnacker on the ground sending disquieting information about delays and processes bogged down by seemingly useless provisions.[43] Yet in the end, their concerns would be justified at least with respect to one issue.

From early on, the Scientific Council had been concerned about provisions in the guidelines that seemed to seriously hamper the prospect of staffing the new agency with appropriate people, and from 'migrating' those working at Directorate S to the new organizational framework. It had insisted that 'the recruitment level of the (scientific) staff and director for the agency' should be high enough to attract highly qualified personnel; 'that scientific staff should also be recruited from outside the Commission'; and that there would be enough staff positions to efficiently run the ERC's demanding decision-making procedure.[44] Commissioner Potočnik had reassured Kafatos that the new agency would have sufficient 'staffing levels' and it would be able to 'offer very good salaries for the very good people we all want the ERC to employ', even though he had to admit that '[w]e are not there yet'.[45] Similarly, Metthey and Cannell reassured the Scientific Council that the transition would be seamless and quick,[46] and tended to brush away concerns of the Scientific Council – 'real problems not expected but it takes time',[47] as the minutes of a meeting briefly after the encounter with Barroso stated. However, in the autumn of 2007 Jack Metthey suddenly spoke of a potential 'crisis': due to the complicated nature of hiring people for the new agency, staffing levels could not be filled quickly enough to accomplish the Scientific Council's goal to run two funding calls in one year for the first time in 2009.[48]

Ultimately, the crisis could be prevented, but more delays took place because skilled staff from Directorate S had formally to re-apply to be taken over by the Executive Agency, and people were lost during the transition.[49] Only by the end of 2008 could Metthey announce that the first recruitments were taking place; otherwise, he was still working on the 'legal basis' (most notably, the Delegation Act, as well as more than a dozen 'Service Level Agreements' with different Commission services and other European institutions).[50] Either the Commission leadership around Silva Rodríguez had underestimated the time to establish the agency, or they did not inform the Scientific Council appropriately.[51] This may have contributed to the Scientific Council's perception that its concerns had not been taken seriously.

But even if Kafatos, Winnacker, and the others felt vindicated about their worst fears and prejudices that there was still 'too much bureaucracy around the ERC',[52] why had the transition to an executive agency become such a nuisance for the Scientific Council leadership? One obvious reason was that, initially, it had hoped that the relative independence of the Executive Agency would mitigate the other frictions that the Scientific Council had witnessed early on; instead, the process had now turned into a seemingly endless series of delays and vexations. But, more importantly, the issue had also been evolving into a power game. At times, at least, Winnacker and Kafatos hoped to convince, or even force the Commission leadership to specifically 'develop a different transition process'[53] in order to transfer staff to the Executive Agency and quickly hire additional personnel, and to speed up the entire establishment of the agency. From his meeting with Barroso Kafatos seemed to have expected that the Commission President would make use of his executive power to overrule Commission protocol in several instances.[54]

If they had succeeded, Kafatos and his colleagues would have proven also more generally that the Scientific Council had some pull on the other ERC entity, and that the ERC compound as a whole was more independent than the legal provisions foresaw. But in the end, the Scientific Council leadership did not achieve its goal. The ERCEA was not established in any significant time shorter than its siblings (measured in the time period between formal legal adoption and operational start of the agency);[55] most significantly, it did not start its operations any time earlier than the Research Executive Agency (REA) that was established during the same period of time, but without the pressure of a group similar to the Scientific Council.[56]

It was no surprise, then, that the period between mid-2007 and mid-2009 was scarred by what Fotis Kafatos perceived as a 'state of crisis' in the internal coordination of the ERC.[57] The Scientific Council and Metthey's headquarters had both come to realize that the ERC compound was not functioning efficiently. Both sides blamed primarily its governance – '*c'est une construction baroque*', as a confidential report of the Commission stated.[58] This brought back with force the initial question whether the actual setup was really suitable to the conduct of the ERC's mission, or whether autonomy would require institutional independence, as the sceptics among the advocates, but also in the European Parliament and in the national ministries had expected.

5.3 Flawed recommendation

As part of the political compromise in imposing the ERC in late 2005, the legal text had stipulated an 'independent review of the ERC's structures and mechanisms',[59] that was now determined to be carried out in 2009.[60] For the Scientific Council, the Mid-Term Review, as the exercise would soon be dubbed, became a kind of point of exit. It was easy to come up with a list of what had been wrong so far: besides the nuisances involving personal matters and the bogged-down implementation of the executive agency, the conduct of the first funding Grant call had also raised 'a number of serious difficulties, ranging from substantial delays to failing to achieve the ERC goals due to stringent regulations'.[61] Yet the review also soon posed a dilemma: should it be used to push for more rights of the Scientific Council within the existing compound of ERC entities, or should the Scientific

Council argue for replacing this framework with one that would make the ERC institutionally independent? Initially, the Secretary General was very much for departure:

> '[S]cientific autonomy cannot be decoupled from control over its operational implementation. This means that ultimately it cannot be effective without complete administrative and financial autonomy. Concomitantly, the ERC must become an autonomous agency, a single, integrated legal entity led by the [Scientific Council, TK], and supported by the DIS as foreseen in the legislation.'[62]

Others were more cautious. In an email to his colleagues in the Scientific Council, Mathias Dewatripont stated,

> I think we should be careful because there are risks in terms of transitions (as we are witnessing with the transition to an EA) and also in the longer term (e.g. if we end up [. . .] with juste retour as an additional mission . . .). Maybe the best strategy is to tell the Commission 'we want to remain with you but you have to help us make the ERC a durable success, otherwise it will be bad for the European scientists and then also for you.'[63]

Eventually, the report by the Scientific Council tried to alleviate both positions. It blamed 'the lack of a clear articulation of a unified institutional structure encompassing the ScC and the nascent Executive Agency,' as well as the 'deficiencies and inflexibilities in the administrative and financial procedures that are obligatory for standard EC Executive Agencies'. It expressed the Scientific Council's conviction 'that the ERC must be a single, operationally integrated science-driven entity', and that it 'sees its future role as that of a Governing Board, and not an Executive Board'.And it proposed 'a pragmatic, two-step approach to more autonomy', with the first step amending the existing framework, and the second step replacing it with a 'novel Community Institution'.[64]

In July 2009, Commissioner Potočnik and the chair of the review panel, Vaira Vīķe-Freiberga, former President of Latvia, presented the results of the review,[65] exciting the international science yellow press, because it was not 'another serving of bland Brussels policy-speak'.[66] Also in content, the Mid-Term Report appeared to be a strong instrument in favour of the Scientific Council's struggle for more autonomy. It detailed, based on a survey of applicants and reviewers, that the ERC funding calls had been 'a huge success, testifying that Europe was indeed in great need of such an instrument beyond the traditional thematic programmes put in place by the European Commission.'[67] Second, it agreed there were 'fundamental problems related to rules and practices regarding the governance, administration and operations of the ERC that are not adapted to the nature of modern "frontier" science management.' Finally, it put forward fourteen recommendations and a roadmap for their implementation, some of which were directed at the Scientific Council (e.g. to professionalize its handling of conflict of interest, and the recruitment of application reviewers), and some at the Commission.

Even though the report did not shy away from strong words – speaking, for example, of 'an incompatibility between the current governance philosophy, administrative rules and practices and the stated goals of the ERC' – it did not recommend changing the current setup. Was the lack of institutional independence a serious impediment for the ERC? Vīķe-Freiberga and her colleagues agreed, in principle:

> In an ideal world, it would have been advisable to set up an independent body run by scientists and managed by scientific officers and staff having experience and expertise in this particular policy area. This autonomy could have been exerted in the framework of an ad hoc institution comparable to similar bodies in national systems.[68]

The difficulty would be to emulate a 'similar body' in the transnational European institutional landscape.[69] The report thus favoured the shelter from political meddling provided by the Commission and suggested alleviating the obvious problems by a few measures overriding the Commission's policy thus far. Only 'if there continue to be major problems over the next two years, [. . .] the organization of the ERC' should be changed 'from the Executive Agency to an article 171 structure'.[70]

Of the recommendations, the most significant concerned 'that the positions of the Secretary-General and Director be merged', and that the position 'should be filled by a distinguished scientist with robust administrative experience.'[71] Soon afterwards, Commissioner Potočnik announced that he would push for amending the legal acts constituting the ERC, and that he would also support the merger. It was the core of his 'two-fold strategy to address not only the classical teething problems [. . .] but also the underlying causes of administrative inefficiencies and structural problems identified in the first period of the ERC's operations', as the Commission's public reaction to the review report had stated.[72] For the Scientific Council, after all the delays, the 'extraordinary speed in changing the present arrangements (and the termination of the function of the Secretary General)' came as a surprise.[73] In its own position paper, it had called for a

'Secretary General jointly recruited by the Commission and the ERC', which could 'even assume the role of Agency Director and become a Commission official' – but, by now attempting to think realistically, it had envisaged such a move only 'in the longer term'.[74]

Recruiting someone from outside the Commission staff for a senior management position was legally possible, but it had not been done before – once routinized, this was a potential gateway for intrusion of national interests into the pan-European services. Thus, pushing it through required some political stamina within the Commission, but with the backing of Vīķe-Freiberga's Report, it could be done.[75] For Potočnik, implementing the merger was probably something on which to build his personal legacy as outgoing Commissioner – a gift to the ERC, proof of his personal care and dedication. To the Scientific Council, it was a sign that the Commission was taking the complaints seriously, and that there was a viable solution to the ERC's governance problem. It had asked to be fully involved in the recruitment procedure, and several of its members were part of the pre-selection panel.

It was also a dream of Kafatos and Winnacker coming true. The former had always thought of the Secretary General as the 'chief executive of the ERC';[76] the latter had explicitly proposed the merger in his draft to the review panel.[77] However, neither of the two would be involved in what happened next. Winnacker's term as Secretary General finished in summer 2009. And in January 2010, Kafatos, due to his failing health, decided to resign from his position as Chair. They were succeeded by Andreu Mas-Colell and Helga Nowotny, respectively. Neither appointment came as a surprise,[78] and they seemed to guarantee that the Scientific Council would reliably continue the track already chosen.

Like Kafatos, Nowotny had been reassured when, in an initial encounter in May 2009, the review panel around Vīķe-Freiberga met the Scientific Council 'with so much understanding and sympathy'. The 'major challenge', Nowotny then stipulated, was not to formulate 'the aims and objectives of what we want to achieve', but rather 'to find the right operational means to get it done without spending another few years waiting for the legislative mills to grind.'[79] Yet Nowotny, soon after assuming her position, dared a complete turnaround. In summer 2010 – the appointment procedure was already well advanced – she wrote a letter to the new Commissioner for research, the Irish Máire Geoghegan-Quinn, urgently pleading her to 'side-step[] the [Mid-Term Report] Committee's well-intentioned, but in its practice flawed recommendation' to merge the Secretary

General and the Director, and asking her to call off the appointment procedure completely.[80]

The merger had been asked for by the Scientific Council; it had been recommended by the Mid-Term Report; and it had been initialized by the European Commission despite the fact that many parts of its services must have perceived it as undermining their own independence. Why would Nowotny want to terminate the one recommendation that seemed, more than anything, to bolster the Scientific Council's position within the ERC compound? Why would she risk returning to the previous situation, which had produced all the problems among the Scientific Council leadership and all the tensions with the Commission? Why would she risk annoying the new Commissioner with this about turn?

One reason was what Nowotny had experienced during the procedure selecting suitable candidates from the 116 applications.[81] Having been published hastily in early December 2009, already the announcement for the open position profile had 'left much to be desired'.[82] As Nowotny wrote to the Commissioner,

> the overall procedure had two important limitations imposed by the Commission rules. One is the age limit of fewer than 61 for applicants, which de facto eliminated at least 50% of eligible good candidates. The second is the fact that the selection procedure strictly followed the internal recruitment rules of the Commission which deviates in some important respects from those practised in the academic world.[83]

However, as Nowotny conceded, 'even under these non-trivial constraints, the three short-listed candidates have the approval of the ScC.'[84] Rather, her decision was based on some far-fetched tactical and strategic considerations mixed with contextual developments over the past twelve months. The most important observation that Nowotny had made was that, since the Executive Agency had become operational, the relationship between the Scientific Council and the administrative branch was improving, and the overall situation was stable. That came with another important change of positions, in spring 2010. Director General Silva Rodríguez moved back to DG Agriculture, and he was succeeded by Robert-Jan Smits. Concluding a breath-takingly successful career inside the Commission services, Smits was an old ally of the Scientific Council leadership, and Nowotny could count on excellent communication with him. Changing the leadership situation in the agency was not pressing any more, and there was also no need to lose the Scientific Council's close confidant and proconsul on the ground – after all,

Winnacker and Mas-Colell had both been selected by the Scientific Council alone.

Nowotny made a simple calculation: the appointment of the new ERCEA Director would bring in someone with no familiarity with the ERC's problems and future challenges. In 2010, the political cycles of the Union were remorselessly moving towards negotiating the next multi-annual budget as well as the next edition of the Framework Programme. Given that the EU was moving toward financial crisis, an experienced administrator from within the Commission would be more valuable to ensure that the ERC would gain a fair share of the next research budget than a new scientist-turned-manager who still had to learn his (or her) ways in Brussels. There was also a strategic component in Nowotny's plan: with the merger, the Commission could claim to have complied with the basic requirements for the Scientific Council, and the option of an institutionally independent organization would probably fade even further.

Eventually, Nowotny's move easily won over everyone (with the notable exception of one of the short-listed candidates for the merged Director-position);[85] but it left the Scientific Council, and the ERC compound, in a strange situation. Never before had the European Commission moved so far beyond its own interests; when the Commission decided in October 2010 to formally 'close this procedure without appointment',[86] Nowotny had declined a unique opportunity to legally enshrine the Scientific Council's own ambition for more power. Instead, she opted for a 'future-oriented and vigorous strategy of finding the best and most durable solution for the ERC governance under FP8, while guaranteeing the full involvement of the ScC in the process.[87] That involved a follow-up on the Mid-Term Report's recommendation to investigate the working of the ERC in another two years' time. In December 2010, the Commission installed yet another expert groups to 'examine the remaining unresolved issues and explore possible governance options to guarantee the long term stability of the ERC structure within the European Research Area'.[88]

While the Mid-Term Report had attempted to maintain the option for an organization independent of the Commission, the 'overriding ambition' of the Task Force Report was to move the ERC 'further into line with international best practice', meaning that the given framework should be further improved. The report came to the conclusion that 'an improved Executive Agency structure is the most appropriate and efficient in the timescale of the Horizon 2020

Programme.' Probably because of its composition (chaired by Smits, it brought together members of the DG Research management, the Scientific Council, as well as others previously involved with the ERC – most notably, Winnacker and Vīķe-Freiberga), the report's language was much more diplomatic and full of complicated euphemisms. It focused exclusively on 'stability' of the existing structure and made more 'specific measures' as to how to achieve 'sustainability and optimization [sic!] of a structure that has largely proven its effectiveness'.[89]

To that end, the report set out, on the one hand, to create 'a cleaner and simpler institutional boundary between the ERC and Commission' and between 'the internal arrangements and in particular the interactions between the two components of the structure (ScC and ERCEA)'. On the other hand, it also argued 'to reinforce the accountability of the ScC and ERCEA', thereby 'considerably relaxing the day-to-day supervision'.[90] Along those two lines, the report made its suggestions well in advance of negotiating the eighth edition of the Framework Programme (aka 'Horizon 2020').[91] But it was left to the Commission to decide whether to take them up (see Figure 5.2).

Appointing two Scientific Council members to the ERCEA Steering Committee, and creating 'a full-time ERC President' – basically a merger between the Scientific Council chair and the secretary general position – were the most significant implementations of better integrating the two entities of the ERC compound. In late 2013, it was all over Brussels that mathematician Jean-Pierre Bourguignon, an ERC Panel Chair of the first funding call, would succeed Nowotny.[92] Other suggestions from the report, however, were omitted once the second tier legislation was drafted for the next edition of the Framework Programme; most notably, that happened to the idea of dispensing the contractual obligations through a 'Memorandum of Understanding' of the ERCEA to the Commission, which would have given more supervision authority to the Steering Committee instead.[93]

With the new edition of the Framework Programme, the ERC compound faced a new roster of regulatory obstacles,[94] and nothing seemed to have changed in respect to the institutional development. There was still a Scientific Council following the fiduciary principle but with little resources of its own, and an Executive Agency strictly committed to following Commission rules; in general, the ERC would remain a funding programme. The only powerful leverage the ERC Scientific Council would call its own was the regular review which, as long as the cycle of political decision-making continued, would

Mid-Term Report (2009)	Task Force Report (2011)	Realization in legislation (2013)
Merge Secretary General and EA Director	Merge Secretary General and Scientific Council chair	Yes: full-time ERC President (as Scientific Council chair)
Director to report directly to Commissioner	Suspense Memoranda of Understanding, '. . . relax day-to-day-supervision by the Commission', 'supervision exercised via the Steering Committee'	No: delegation act and Memoranda of Understanding implemented
Fair composition of EA Steering Committee: two members of Scientific Council, and one outside distinguished scientist	Two members of Scientific Council to be members of Steering Committee	Yes: two members of Scientific Council are members of Steering Committee
Compensation for chair and vice chair in the form of lump sum	'New mechanism' for the support of the chair and the vice chairs	No: annual support action for vice chairs
Another review in two years	'. . . the mid-term of the Horizon 2020 programme'	Routinely scheduled mid-term review of Horizon 2020
'. . . to change organization to article 171 structure' in time of the next edition of the Framework Programme	'. . . may be necessary to rethink the structure, including a renewed consideration of the possibilities offered by Art.182(5) of the Treaty'	Not mentioned
Board 'an extremely useful liaison instrument between the scientific and the management components'	Make Board legal and call it 'coordination committee'	Yes: 'regular coordination meetings'

Figure 5.2 Suggestions on changes in ERC governance structure

Compiled by the author.

still be an integral part of it. But a few things had changed. The two entities could depend greatly on the social fabric created at the premises of the Executive Agency. And, most notably, the ERC was now outfitted with an impeccable reputation across Europe and the world.

— 6 —

A RATHER CONVENTIONAL SYSTEM

The deadline for the first ERC funding call was in May 2007: internal projections had expected up to 6,000 applications,[1] yet the call yielded a 'massive 9,167 proposals'. The press release tried to diminish it as 'impressive demand',[2] but for a short time, the hefty oversubscription overwhelmed both the Scientific Council and the small team in Directorate S handling the proposals and administering the selection procedure. The crisis could be overcome, thanks to '[v]ery enthusiastic and dedicated staff';[3] the fact that the Scientific Council members got on the phone to organize about 600 additional panel reviewers; and that national funding agencies sent staff to help out.[4] The Scientific Council quickly reassumed its position as guardian of the integrity of the selection procedure,[5] but the experience would leave scars.

Between 2007 and 2013, the ERC was tasked to allocate more than € 7 billion to research projects. This sum alone would hardly explain why the new funding instrument was gaining so much attraction. But the generous grants that the ERC offered were expected to come along with something exclusively symbolic. Rather modestly, Keith Pavitt had called it 'high-quality academic research';[6] EURAB later just happened to arrange the buzzwords most succinctly: the ERC was destined to 'support world-class research [. . .] by adequately funding scientific excellence through competition at a European level in fundamental research.'[7] Every researcher with ERC funding would be recognized and pride herself on being awarded an academic accolade that was above and beyond national recognition.

To tie monetary value to symbolic distinction, the allocation of funds would have to rely on specific tracks and schemes that would define, among other things, who would be eligible to apply, how much money would be available per project, and what the expected research should be proposed. The various schemes would translate the ERC's overall mission to actual funding opportunities for researchers. However, even more importantly would be to implement a procedure of decision-making, that is, a procedure determining which applications to select for funding.

6.1 A broader palette

One of the most exciting exercises in the early meetings of ELSF and ISE had been to dream up the ERC's funding opportunities. Thus, those meetings, under the command of Frank Gannon and skilfully set in place by Luc van Dyck, provided a lot of thinking;[8] some aspects had been covered by the Commission, such as the fact that the ERC should make contributions in the form of grants.[9] Soon after the first Scientific Council meeting, Fotis Kafatos, supported in London by Babis Savakis, as well as Helga Nowotny and Daniel Estève, summed up the list of features that the future ERC funding opportunities should contain, as well as features that should be avoided. Other good resolutions included that a fair amount of the ERC's funding budget should be dedicated to 'young investigators and newly established teams', with the option of 'portability to give greater opportunities to the young researchers'; that the ERC should not 'create an administrative monster';[10] that the 'selection process' should rely on panels, 'with a disciplinary cluster as a core';[11] and, ultimately, that the ERC's mission was defined as 'stimulating investigator-initiated frontier research across all fields of research on the basis of excellence'. It was a great phrase, but the tricky part was to stitch together a coherent 'scientific strategy'.[12]

Based on the initial papers, William Cannell drafted an 'outline strategy note', bringing together the scattered ideas in order to determine 'the main parameters for the ERC's scientific work programme for the start of its operations'. As the paper stated, the ERC

> must operate at the highest level of ambition to generate the maximum benefit to European research from the activities it pursues. As a new organization, it should not be hostage to the conventional wisdom; instead, it should take the best practice wherever it can be found, and should welcome 'positive contamination' effects on the European research system.[13]

What would be the best way of making sure that the ERC would not be diverted and stay true to its ambition? Cannell's paper stressed the need for 'simple procedures that maintain the focus on excellence, encourage initiative and combine flexibility with accountability'. It demanded that the ERC 'should also complement and not duplicate other research activities in the 7th framework programme' – in other words, to focus on its niche in the already richly populated funding landscape. The note had the full sympathy of the Scientific Council, as it successfully formulated an emerging consensus within the body: keep it simple.

Cannell's ingenuity in taking up and re-packaging the discussions and contributions made earlier by Scientific Council members can be illustrated by two examples. In her position paper, Helga Nowotny had envisioned five different funding streams, one on people and one on projects, and each of them split for younger and for senior scholars; in addition, she also suggested a dedicated scheme for interdisciplinary projects.[14] While maintaining the distinction in two separate tracks for researchers at different stages of their academic career, Cannell relocated that between people and projects into the evaluation criteria, which he foresaw as 'a blend of the potential of the people and their track record' and the 'excellence of the project'.[15] Similarly, Cannell also solved how to determine the notion of 'young', or 'early-stage' researchers without being discriminatory against those having an uncommon career. Daniel Estève had asked for 'a set of simple common rules'.[16] Learning from the EURYI calls, Cannell's paper took the academic age as starting point, meaning the timespan between the award of the PhD and the time of submitting a proposal, thereby ignoring the biological age.[17]

In short, the ERC funding portfolio was devised by William Cannell under the supervision of the Scientific Council with the ambition to offer highly attractive conditions to researchers. There would be two schemes: one to early-stage career scientists and one to senior scientists; both would run annual calls; each would offer one-size grants with enough funding to support project leaders (in research funding lingo, the 'Principal Investigator', or PI) and their teams for up to five years; applicants from all over the world would be eligible to apply; a grant would cover a substantial overhead of 20 per cent and it would be portable with the person scientifically responsible. Most significantly, the funding would be 'bottom-up', by which was meant that there would be no predefined topics along which proposals should be oriented. Instead, the ERC promised to be capable of identifying the best submissions, whatever area or field or new question they proposed to tackle.

How well Cannell mastered the arrangement of the Scientific Council's scattered ideas into a coherent structure could be seen from the fact that the group amended only very few, and – from a general perspective – rather cosmetic issues of his strategy note. In particular, it replaced the bureaucratic abbreviations used so far for the two funding streams, namely 'Early Stage Independent Investigators' with 'ERC Starting Grant', and 'Established Investigators' with 'ERC Advanced Grant', respectively.[18] By April 2006, the ERC had its 'scientific strategy' lashed together in broad strokes:

> It is envisaged that these two funding streams, operating on a 'bottom-up' basis, across all research fields, should be the core of the ERC's operations for the duration of the 7th Framework Programme.[19]

Over the next decade, the ERC funding portfolio would remain surprisingly stable in the sense that the core concept of two funding tracks would remain in place. That didn't necessarily mean that there were no modifications to the ERC's established funding portfolio: not only was the conceptual design of the respective funding stream routinely revisited, such as financial endowment and eligibility criteria; each of the two funding tracks would also see significant changes (see Figure 6.1). From 2013 on, the early-stage career track was split in two funding schemes: the 'Starting Grant' scheme's eligibility window would be narrowed (two to seven years past their PhD), and it would be accompanied by the 'Consolidator Grant' (with applicants eligible from seven to twelve years past their PhD). The reason for this split was that the demand for this funding track was beginning to overwhelm the decision-making procedure, and that the Scientific Council feared that researchers closer to their PhD (and, thereby, 'younger' in terms of academic achievements) would lose out to their more advanced peers. Except for the different eligibility windows, the two schemes would complement the early-stage career scientists' track.[20]

A couple of years into operations, the Scientific Council also started to discuss whether to introduce new funding opportunities. The initial rationale behind the brainstorming was primarily a strategic one, namely that the ERC should extend its realm of action; as Andreu Mas-Colell, the second Secretary General, held in a memo written together with staff from the Secretariat:

> The NSF has 300 programs. This is too much, but two, Starting and Advanced Grants, is too little, a broader palette will make the case for the ERC most compelling.[21]

Track	2007	2008	2009	2010	2011	2012	2013	2014
Early-stage career scientists	StG		StG	StG	StG	StG	StG	StG
							CoG	CoG
Senior scientists		AdG	AdG	AdG	AdG	AdG	AdG	AdG
						SyG	SyG	

Figure 6.1 ERC major funding calls, 2007–2014

Compiled by the author; StG = Starting Grant; CoG = Consolidator Grant; AdG = Advanced Grant; SyG = Synergy Grant.

For a while, the Scientific Council was tossing around different ideas, such as turning to institutional excellence and to fund one or several 'Institutes of Advanced Studies', or to introduce 'twinning grants' to Principal Investigators in order to support researchers from so-called weaker-performing countries.[22] Additional funding opportunities were difficult to align with the principle of simplicity that had, by now, become an important fixture among Scientific Council members. In the end, however, the Scientific Council decided to add a new funding stream in relation to the senior scientists' track. Initially dubbed 'ERC plus', and later called 'Synergy Grant', its basic idea was to award an even bigger chunk of money to two to four PIs of equally excellent format for a large research undertaking in order to support new, interdisciplinary approaches. It could also be interpreted as a subtle concession that, in the perception of some members of the Scientific Council, the 'Advanced Grant' scheme was not really fulfilling its intended purpose, namely to stimulate senior scientists to propose risky and unprecedented research proposals that they would not have embarked on otherwise.

In some respect, the 'Synergy Grant' scheme might have been the logical successor of the 'Advanced Grant' scheme; yet as ambitious as it may have been, the new scheme also deviated from the ERC's own, rather narrowly defined philosophy centring on individual Principal Investigators, and it also extended the meaning of excellence, since it included not only the personal capacity of one person, but asked for some collaborative aspects as well.[23] Due to this feature, there was the fear that it might be confused with collaborative research funding provided by other parts of the Framework Programme, and hence it was not well regarded by several members within the Scientific Council as well; its introduction was primarily due to the forceful execution and subtle arguments by the group's chair,[24] and it would only run as a pilot for two consecutive calls in 2012 and 2013.

One key problem of the new scheme was that the involved reviewers lacked a clear understanding of what was meant by 'synergy'. Even though the Scientific Council tried to identify peers with a very broad realm of expertise, many proposals were in danger of not being covered. The first call on the new scheme was thus marked by a lack of clear 'communication of what "synergy" is about'.[25] That pointed to a larger issue: the mismatch between the ambition of this scheme and what could actually be achieved. As important as the funding portfolio was to transfer the ERC's mission into distributing funds, even more crucial was that the decision-making procedure would be commonly and convincingly perceived as fair, reliable, and funding the proper proposals. On getting this obligation right would depend the entire credibility of the ERC.

6.2 To promote interdisciplinarity and breadth of viewpoints

Theoretically, there would have been several different ways of how to allocate its available budget to research,[26] yet for the Scientific Council, there was little doubt that it wanted to base its funding decisions on one principle: peer review. Ever since it was first introduced to research funding, peer review had been an unabated success, due to the fact that, unlike any other mode of distribution, it achieved dual legitimacy among applicants – many of whom necessarily questioned why they were not funded – as well as among politicians – who, after all, had been allotting the funds in the first place.[27] Peer review, in the most general meaning, primarily referred to a decision-making principle, i.e. to base a funding decision on a proposal's assessment by informed peers. To realize this principle, however, it had to be implemented through a procedure, which had to organize at least four components to define and prescribe 'step by step' the 'processes of negotiation, selection, and verification' of submitted proposals, leading reliably 'to some kind of outcome':[28] the process-flow determining the consecutive steps of decision-making, to ensure a robust and transparent procedure; the criteria to accord with the specific funding scheme, to assure that proposals were assessed against strategic intentions; the peers selected for reviewing, to judge submitted proposals; and the methodology of assessing the proposals, to arrive at a clear choice at the end of the procedure.

While peer review quickly became the most prominent (and most reputable) mode of distributing funds to research projects, it was surrounded by doubts about its integrity,[29] which have only increased with the rising importance of competitive funding for a successful academic career, and a general shift in science policy towards stifling more competition within the academic market;[30] as a consequence, peer review's complexion profoundly changed over the past decades.[31] The doubts addressed different points of critique, each dealing with inconsistencies between the principle's idealistic assumptions and the reality within which the procedure was implemented.

To start with, the idealistic assumption that everyone involved would act solely in favour of scientific qualities was inconsistent with the simple fact that scientists, as human beings, were driven by other, more worldly ambitions than just intrinsic values, too.[32] Scientific malpractice in the broadest sense included not only deliberate machinations and wrong-doing by individuals,[33] but also seeping suspicions about common prejudices and cultural stereotypes that would not be smoothed out, but rather reinforced through the decision-making procedure.[34]

Another inconsistency was due to the fact that peer review was actually a very time-intensive activity; it was deemed efficient partly because the costs for producing the input were mostly delegated to the realm of general academic activities. An applicant was expected to write up her proposal either during her regular employment or in extra shifts, and reviewers were often expected to assess proposals with no further costs. Yet with both the absolute number of peer review procedures and of submissions increasing, there was widespread feeling among scientists that they had to dedicate more and more of their time to either writing proposals or assessing them.[35] It was becoming doubtful if, overall, the individual investment of time was still in a good relation to what was actually achieved.

The final inconsistency was that, while the principle seemed to assure that the best proposals were selected, deciding what was 'the best' depended very much on the reviewers' judgement; in the assessment of submitted proposals, promise and feasibility were in opposition to each other: if the former tried to push the research into new frontiers, the latter was to reduce the 'risk of getting entangled in a promising but not feasible and/or not relevant proposal'.[36] Thus, peer review had long been suspected of being in support of mainstream rather than 'unorthodox or high-risk research'.[37] For the ERC, the simmering reservation in many corners of the academic world that peer review would be an inherently conservative system was probably the severest form of doubt.[38]

As a group of highly esteemed scientists, many of whom had long-standing experience in science policy, the Scientific Council was fully aware of those doubts. Yet they were not perceived as a constraint, and definitely not as a reason to abandon peer review altogether; rather, they posed a quest for composing the ERC-specific implementation of the procedure in a way that they would be rebutted as convincingly as possible. As Kafatos maintained, the ERC's 'focus [. . .] on top quality, leading edge, innovative research [. . .] mandates a certain innovativeness in the structure we choose'.[39] From early on, discussions revolved around how to counter, or outflank, doubts, with the ambition for the resulting ERC-specific implementation of the peer review procedure to win full acceptance among scientists and policy-makers alike. The resulting procedure (see Figure 6.2) was first put to the test with the 2007 Starting Grant call.

To define a process flow, the Scientific Council had come up with a generic division in three domains[40] with twenty panels, eight in (what was soon called) the Physics and Engineering (PE) domain, seven in Life Sciences (LS), and five in Social Sciences and Humanities (SH).[41] Another item concerned the sequence of procedural steps: reviewers should come together to jointly decide on the selection of proposals in two panel meetings, first for preselecting from all proposals those that would deserve closer scrutiny, and second for determining the list of proposals that were actually to be funded. Preparing for each meeting, the submitted proposals would be checked for eligibility and then assessed by individual reviewers. The first panel meeting would discuss the entirety of proposals, retaining a certain number for more detailed evaluation; the second meeting would then provide room for interviews with the remaining applicants and conclude on the rank list of projects to be funded.[42]

The second component concerned the review criteria that defined the lines along which submitted proposals should be assessed. The legal text had simply provided that 'projects selected for funding must demonstrate a high scientific and/or technical quality';[43] the Scientific Council maintained that reviewers should be exclusively asked to assess a proposal's scientifically intrinsic values, since 'excellence is the basis for evaluation'.[44] When the Secretariat proposed to include, among a set of questions (in ERC lingo, called 'elements') to be considered by the reviewers assessing proposals, to also ask for the 'potential impact' of the research proposal 'for innovation and social/economic benefits',[45] the Scientific Council members reacted almost allergically. Impact would be 'in many

Components	Operational items	Specific ERC implementation 2007	Routinized amendments/ substantial changes
Process flow	Topical	3 domains; 20 panels	Refining panel descriptors/ 2008: No. of panels extended to 25; 2008–2011: fourth domain on interdisciplinary proposals
	Sequential	Individual assessments, 2 panel meetings (early-stage career track: Interviews)	None/ None
Criteria	Values targeted	Intrinsic scientific values	None/ None
	Objects of assessment	Proposal; PI; research environment	None/ 2012: 'research environment' removed
Peers	Selection	'Hand-picked'	Appointing new PMs/ None
	Roles	Panel chairs; panel members; remote reviewers	None/ None
Methodology	Relating	Scoring and ranking	Refining scores/ None
	Decision	Autonomous by panels	None/ None

Figure 6.2 ERC implementation and amendments of peer review core components

Compiled by the author.

cases too difficult to determine;'[46] the question reformulated asking whether 'the research [would] open new and important, scientific, technological or scholarly horizons'.[47] In addition, the Scientific Council determined that the reviewers should be asked to assess three specific objects (in ERC lingo, 'headings'), namely the project proposal, the CV of the Principal Investigator, and the research environment in which the project should be carried out.

As for the third component, the Scientific Council distinguished between three roles for peer reviewers: remote referees, panel members, and panel chairs. The main focus was on panel members: they were expected to have a broad expertise, a good overview of more than one research field and, if possible, also would cover more than one academic discipline. Most importantly, they would have to have a good reputation and to be as blind to potential biases as possible. Panel chairs were devised as a distinct group within the panel members; as primus inter pares, a panel chair was expected to be a particularly highly esteemed and also trusted scientist. To complement expertise where necessary during the procedure, remote reviewers would be asked to assess individual proposals in preparation for the second panel meeting.

It was well known among Scientific Council members that, in order to get the peer review procedure right, it would be crucial to establish a certain 'panel culture' among the panellists.[48] To that end, the proper scientists had to be selected, and this was deemed such a sensitive issue that, when the exercise was done for the first time in 2006, the Scientific Council ignored existing lists that were provided by the European Commission or other funding organizations, and that sometimes were based on self-nominations.[49] Instead, members of the ERC review panels should be 'hand-picked' by Scientific Council members,[50] and, in order to assure 'a balanced and inclusive distribution', the Secretariat would work on the 'final composition' of a panel.[51] As they were of secondary importance, the remote reviewers were to be suggested by the panel members as they saw fit.

Finally, the methodology of the procedure was based on a traditional scoring system, which would, first, allow individual reviewers to grade different proposals and, thereby, let the group of panel members transfer the qualitative assessment of proposals into an easily comparable range of numbers for ranking. Scoring and ranking was a necessary instrument as a panel would have to identify quickly those proposals that were deemed to be of highest quality from among a list of up to more than 200 submissions. Yet the Scientific Council also 'stressed the importance of coherent interpretation of the evaluation criteria by the different referees, panel members and panels';[52] to achieve this, ultimately, decision-making would remain the extraordinary remit of a panel and it would be allowed to overrule any individual score as it saw fit. The Scientific Council also prevented itself from making any adjustments to the ranked list of proposals coming from the panels.[53]

Overall, the ERC-specific implementation of the peer review

procedure turned out to be 'a rather conventional, nationally well tested, system with a panel of experts with knowledge in depth of the field', as an external observer would later state dryly.[54] In order to appear fair and to base decision-making on impartial yet informed assessment of submitted proposals, the Scientific Council strove to set up international and diverse panels, composed of scientists of highest acclaim. To avoid the appearance of inefficiency, the panel members were paid an honorarium, and the process flow was made as smooth as possible; in addition, proposals should be short and concise, and anyone fulfilling the two-to-nine years past PhD criterion should be eligible to apply. And to nudge reviewers to go for risky, non-conservative proposals, the Scientific Council had not only written the questions to the reviewers accordingly, but also created a panel structure that it hoped would 'promote interdisciplinarity and breadth of viewpoints within each panel.'[55]

6.3 Preventive and dissuasive actions

The Scientific Council members would soon pride themselves in having created a 'pan-European peer-review system'[56] that was 'truly international with peers from the entire continent and overseas',[57] and to ensure that 'quality and originality of the research project and the qualifications of the applicant [. . .] are the only evaluation criteria.'[58] It was a 'gold standard' for Europe,[59] implying that the ERC was leading the way with how to conduct properly the allocation of public funds along the principle of peer review. Indeed, the resulting ERC-specific implementation of the peer review procedure successfully managed to juggle three different requirements: it reflected the strategic mission of the ERC to fund 'frontier research', that is, 'inter- and multidisciplinary, high risk pioneering projects and new groups and new generation researchers';[60] it turned out to be reliably stable, even in cases like the oversubscription in 2007, internal glitches, or external havoc;[61] and, overall, it would be perceived as fair and balanced to everyone involved.[62]

This did not mean, however, that the doubts about peer review in research funding would cease, or that the ERC-specific implementation was beyond critique. The procedure would have to be continuously orchestrated[63] – as the results of the 2007 Starting Grant would indicate, that included not only to maintain the 'operational' integrity of the procedure – to 'manage the workload of proposals in a way that maintains quality and user-friendliness within a reasonable time frame for the evaluations,' as an internal report called it. Equally important, the same report continued, was to keep an eye on the 'political' implications, meaning that

> the impact of high demand – deriving from possible operational issues and also from the impact on applicants (low success rates) and peer reviewers (high workload) – [might] lead to dissatisfaction with the ERC among the stakeholders.[64]

The Scientific Council readily accepted the exercise to constantly refine the operational items of the ERC's peer review procedure, and it understood that this was not only to make the procedure watertight,[65] but also to mediate expectations to different audiences. At its core, it built on two complementary messages. One was to assign 'excellence' to be the key role in the procedure.[66] The strict emphasis only made sense in the context of European research funding, where peer review had been in place for a long time, but where other criteria next to scientific quality had been added as well.[67] The second message was the key phrase of the ERC being a 'learning organization'.[68] This required the Scientific Council and the ERCEA management to establish a tight control over different aspects of ERC's decision-making machinery, closely monitoring the actual conduct of funding calls and implemented instruments such as indicative call statistics, panel observations, and panel reports.[69]

The most common strand of orchestration activities concerned routinized amendments to some of the items of the procedure; most notably, those concerned panel composition, refining the scores, and revising the panel descriptors. Between 2007 and 2013, the Scientific Council reformulated the so-called 'descriptors' that were established to indicate which academic disciplines and research fields were to be covered in a panel, and over the years, this exercise led to a complete reorganization of some panels. Panel members were replaced on a continuous basis, not only to bring in new faces, but also because of new weighting of the panel's expertise, and because some panellists had not been performing well. Finally, the methodology of marking, scoring, and ranking the proposals was regularly revised, and extended, thereby not only quadrupling the text size on that specific matter, but also making it more complex and nuanced. The primary instrument for routinized amendments was the Work Programme, a legal document used across the FP format for a certain period (usually one year) to allocate a budget for specific funding streams and define the means and operative terms to be followed; it was here that the Scientific Council crafted the conceptual meaning of excellence and translated it into operational means.[70]

As important as they might have been, the routinized amendments and changes and the setting of additional rules alone were hardly enough to do away with potential misbehaviour and unfavourable perceptions. The Mid-Term Report had tasked the Scientific Council to set up two standing committees, one 'to steer and control the construction of a database for the selection of reviewers and panellists' and the other 'dealing with conflict of interest issues'.[71] In addition, the group also convened several ad-hoc working groups; most notably, one was dealing with 'Gender Balance', and another with 'Widening Participation'. Their role was to focus on topics that were deemed potentially detrimental to the reputation of the ERC's peer review procedure, and to engage in secondary activities such as sponsoring studies about the fairness of the procedure, establishing routines of providing feedback to rejected applicants, overhauling and extending the fabric of rules and procedures for applicants and reviewers, and publishing exhaustive statistics on funding calls and success rates, to mention just the most obvious ones.

A common and tenacious suspicion of a collective bias related to gender. It was not so surprising that fewer women scientists applied for ERC grants (and particularly for the 'Advanced Grant' scheme), given the fact that the share of women in the top tier of the scientific professions was very low across all European countries.[72] What was concerning was that the share of those female scientists successful with their ERC application was lower than the initial share of women among applicants.[73] To tackle the issue in the allocation of its funds, the Scientific Council adopted the 'Gender Equality Plan', consisting basically of a number of soft measures 'with the focus kept on excellence', promising to challenge '[a]ny potential sources of gender bias in the evaluation process', and, as 'a medium term goal', to 'achieve gender balance in each ERC evaluation panel as well as among the panel chairs.'[74]

The 'plan' hardly meshed with the actual implementation, yet it was received with great sympathy; quickly, the Scientific Council embarked on drafting similar reports on other topics potentially detrimental to the integrity of its decision-making procedure. When it published its 'Scientific Misconduct Strategy' in late 2012, the Scientific Council had, for a while already, discussed the difficulties of how to deal with applicants that were detected having engaged in fraudulent behaviour. Unlike national funding instruments for academic research that have put tough measures in place to counter scientific fraud,[75] the ERC had no sanctioning power: it could not, for example, debar a PI or a host institution from future funding calls. Nor could it simply issue name-blaming reports. And even higher authorities could not be involved.[76] The adopted paper graciously tried to cover its weakness.

> The ERC strategy is fundamentally based on the presumption that the host institutions of the ERC applicants and grant holders have the primary responsibility for the detection of scientific misconduct and for the investigation, and adjudication of any breaches of research integrity that may arise.[77]

Yet the Scientific Council leadership must have felt that this would not be enough, and Helga Nowotny and Pavel Exner told the following story in an article in *Science Magazine*:

> Imagine sitting over a pile of applications submitted to one of the most prestigious funding agencies. Suddenly, what you read appears familiar—not only the idea, but its terminology and the methods proposed. You recognize entire sentences because you wrote them.[78]

Had not the (anonymized) panel member been assigned to review the proposal that she then identified as the same text that she herself had submitted to a funding agency in a different part of the world some years ago, the issue would not have been detected at all. Besides a good story about appalling behaviour of a distinguished scientist, Nowotny and Exner calculated that this open statement would also signify the Scientific Council's best intentions to protect the ERC procedure.

In a similar vein, the Scientific Council's ambition to present the peer review procedure as fair and clean was vulnerable to the danger of inadequate behaviour of peers – the very people expected to look at proposals impartially and coolly while, at the same time, being energized and embracing unconventional ideas and thinking outside the box. To prevent hidden agendas of peers, the European Commission had provided for a seemingly 'clear set of rules pertaining to conflict of interest' in the FP format's legal specifications.[79] Those rules distinguished between a 'disqualifying' and a 'potential' conflict of interest; in the first case, which included all sorts of personal or professional relations to an applicant, the reviewer 'shall then neither assist in the individual assessment [. . .], nor speak [. . .] in any Panel discussion related to this proposal'. The latter was merely a safety measure for 'cases not covered by the clear disqualifying conflicts'; here it would be up to the administrators on the ground to 'consider the circumstances of the case and make a decision'.[80]

The measures to rule out conflicts of interest had their own effect on the peer review procedure. For one, the rules held that, in case a panel member 'is' or 'was employed by one of the applicant legal entities [i.e. the applicant's host institution, TK] in a proposal within the previous three years', he or she would be 'disqualified' from reviewing and discussing that specific proposal.[81] For panel members from France, where many researchers were affiliated with the country's sprawling national research organization (Centre national de la recherche scientifique, CNRS) it meant that they were automatically excluded from almost all applications from that country (a similar case, although not so severe, was of the German Max Planck Gesellschaft).[82] At the same time, the Scientific Council also witnessed different understandings and interpretations about how seriously reviewers of different cultural and disciplinary background would take the issue.[83]

As much as the ERC Scientific Council made use of routinized amendments, of public statements and other 'preventive and dissuasive actions',[84] it was wary of touching the core setup of its overall implementation of the peer review procedure. Actually, there were only two substantial changes between 2007 and 2013 (see also Figure 6.2, last column to the right). One concerned altering the objects of a proposal that were to be assessed by reviewers; the other concerned the topical structure of the process flow. Both were, again, the result of the need to improve the integrity of the procedure as much as giving a clear statement of intent to the public.

Initially, and up until 2010, the Scientific Council had set out the 'sole criterion of excellence' on three headings – the PI's career, the project description, the research environment.[85] The latter was supposed to ensure the feasibility of a research project; yet upon closer analysis, it was concluded that this heading 'has been understood and used – in some cases clearly misused – in many different ways'.[86] Consequently, the Scientific Council decided to do away with it and to restrict reviewers entirely to assessing applicant and project outline.

At first sight, the decision seemed to be straightforward: if a component of the set of questions turned out to be too ambivalent, it would be better to drop it. This interpretation conformed to the self-image of the ERC that took its ambition to be perceived as making everything simple and transparent. But it also evinced an eminent political concern: assessing the research environment was perceived as having a potentially negative impact on an otherwise good-enough proposal because of a weak institution. By the time the issue was discussed in the Scientific Council, the regionally imbalanced distribution of ERC grants could not be disguised any more.[87] All the more, there was no reason to fuel the irritation of scientific communities (and their political representatives) in some of the 'weaker' countries. The Scientific Council's answer to this specific problem was seemingly bold, but actually very easy: the issue of the research environment could also be evaluated under the heading concerned with the project proposal,[88] but with much less political exposure; at the same time, the ERC would only gain in standing, as scrapping the heading 'might also send a signal on the openness of the ERC to proposals from any [host institution, TK].'[89]

This was a fine example how the Scientific Council used the Work Programme and its section on the evaluation criteria to communicate proactively – not only with its reviewers, but also with potential applicants, management of research institutions, and policy-makers about the prevailing fairness of the procedure. But changing the evaluation questions alone would not suffice. Here, as well as in other instances, the Scientific Council recognized that it would have to send a regular message to its stakeholders that it was honestly engaged in tackling the unequal distribution of its grants. In a similar way to the provision in the 'Gender Equality Plan', altering the composition of the panels in favour of more reviewers from countries with less than average ERC grants seemed to be the most reassuring and comforting strategy.

As much as the review questions aimed at nudging the applicants and the reviewers towards more risk-taking, the Scientific Council remained concerned about the panels being really able to identify truly outstanding proposals. At first, this focused on the question of how to treat proposals fairly that, by definition, would go beyond the border of a given panel. The idea of broad panels had been a striking one, and one of the few truly innovative features of the ERC's implementation of the peer review procedure; it was understandable that the Scientific Council, after raising the number of panels to twenty-five (ten in PE, nine in LS, six in SH) in 2008,[90] would not accept any further extensions in order to avoid diluting its overall ambition to have broad, interdisciplinary expertise gathered in one room. Inevitably, however, there would also be some proposals across the borders of the particular expertise combined in any given panel. Given the fact that research ideas could develop in new and unexpected directions, the issue of proposals dealing with topics covered in two (or even more) different panels was of great importance to the Scientific Council; not to mention that such 'cross-panel' or even 'cross-domain' proposals were expected to be of particularly high potential.

However, there was a dual problem, one being terminological, and the other being procedural. The procedural problem was, of course, how to deal with those cross-panel proposals. The initial idea was that proposals not entirely covered by a panel should be flagged and then be assigned to the so-called 'fourth domain', a separate budget line from within the overall funding for a given funding call; the proposals in this pot would then be presented to the panel chairs at their final meeting.[91] And further:

> taking account of the forward looking and innovative nature of the programme, all the peer review evaluation panel chairs or their deputies will bring forth and specifically discuss, from an interdisciplinary perspective, the scientific added value of proposals above the quality threshold which are of interdisciplinary nature. In order to establish the ranked list of the Interdisciplinary Research domain, all peer review evaluation panel chairs will further assess these proposals on the basis of the second evaluation criterion (Research project).[92]

As it would turn out, the additional step was not very effective. The allocation of funds was usually done 'almost entirely between the two Chairs involved', since the other chairs, obviously from very different fields, 'had not read the applications'.[93] Ultimately, the Scientific Council decided to abolish the fourth domain effective with the *Work Programme 2012*; instead, the cross-panel proposals would now be mainstreamed during the regular evaluation procedure, and their ranking would 'simply follow the administratively "fair" approach which is currently used as the starting point of the discussion'.[94] The problem was not solved, as an analysis later would show, because 'the a priori success rate of "ID proposals" [i.e.: cross-panel proposals] is clearly lower than that of non – ID'.[95]

The quote already indicated the second, terminological, problem that the Scientific Council was facing. To use the paraphrase 'interdisciplinary' for those proposals was actually misleading, given the fact that the panels themselves were also composed of experts from different (neighbouring) disciplines, and could well deal with interdisciplinary proposals, as long as those coincided with the expertise in the panel. The misconception was not easy to get rid of, neither in communication with the research community and potential applicants, nor (obviously) in the internal analysis. It was regularly leading to confusion that cross-panel proposals would have to be particularly risky and outstanding. The Scientific Council did its best to challenge this impression: actually, it wanted to avoid any qualitative distinction between proposals covered fully by a specific panel and transverse proposals. 'We want everybody to go for high risk', as Kafatos noted from a discussion.[96] After a while, the issue seemed to be less pressing to the Scientific Council. The overall peer review procedure was performing very well, and the burden of proving that the ERC was funding ground-breaking research was eased by the overall success that was assigned to it.[97]

— 7 —

WIDE-RANGING EFFECTS

In early October 2010 the ERC announced that one of that year's awardees of the Nobel Prize in Physics, Konstantin Novoselov, had also been granted ERC money a couple of years ago.[1] The press release came on the same day as the Swedish Foundation presented its newest Laureates; it was an annual routine that, reliably, would create a media buzz. The ERC, however, could not claim that its funding had enabled Novoselov the research for which he had now been selected to step into the illustrious group of Nobel Laureates. The path-breaking discovery had occurred years earlier, when Novoselov and his colleague and mentor, Andre Geim, had found a way to produce 'atomically thin carbon films' and began investigating their surprising physical attributes.[2] By the time Novoselov submitted his successful ERC project proposal, 'graphene', as this mono-layered substance came to be known by then, 'had become a gold mine for searching for new phenomena'.[3]

Why, then, was the ERC riding on the Nobel Prize's coat-tails? One reason was that Novoselov had been one of the applicants for the highly oversubscribed 2007 'Starting Grant' call – to the ERC leadership, the Nobel Foundation's choice was seen as a retroactive confirmation that the ERC panels had, among thousands of applications, picked just the right ones: the decision-making machinery was working. The ERC public press release, however, made a different argument. It called the Nobel Prize 'a recognition of the type of work funded by the ERC, focussed on research at the frontier of knowledge'.[4] Indeed, Novoselov's work seemed to nicely bring together what the notion of 'frontier research' intended to embrace – pure curiosity and insistence leading to radically new insights, which in turn promised to positively affect society and, more importantly, stimulate economic growth.

The Nobel Prize also gave the ERC a welcome recognition from the political world. In the ERC press release, Commissioner Geoghegan-Quinn was quoted as applauding Novoselov and Geim, as well as the ERC. She added her expectation that 'more Nobel Prizes will follow as a result of this valuable European investment in the best scientists and in their innovative research in Europe.'[5] The ERC understood the message: it dutifully continued to report grantees who became Nobel Laureates.[6] The media attention borrowed from Nobel Prizes, however, would not suffice; ultimately, an R&D funding instrument like the ERC had to provide evidence that it was fulfilling its objectives and that it was contributing to the overarching policy goal, that is, innovation. Since the ERC had been fashioned to be a new path in research funding in the context of the Lisbon Strategy, it would yet have to establish its own means of proving its value; as it turned out, there was a lot of leeway in the interpretation of objectives and also in the provision of evidence. Yet, developing a methodology of evaluation criteria, conceptualizing the findings within the broader political framework, and reacting towards emerging policy trends would keep the Scientific Council and the Secretariat busy.

7.1 Justifying to stakeholders

Impact assessment has been a proliferating field in R&D policy for a long time, with different approaches and methodologies (benchmarking, comparison, case studies, scientometrics). Yet, as diverse as the approaches may have been, they all shared two commonalities: while they were based on scientifically sound evidence, their common set of denominators (such as innovation, basic or frontier research, and interdisciplinarity, to name but the most common ones) remained ambiguous. It should not come as a surprise, then, that the evaluation of funding programmes within the FP format usually served political purposes rather than objective targets.[7] Assessing the ERC's performance, too, would provide opportunity along the same lines.

Discussions about the methodological apparatus as well as the concept of what should actually be assessed had been going on between Scientific Council and Secretariat since early on. They were based on the realistic assumption that, within a few years, the ERC would have funded several thousand projects, the sum of which would, theoretically, provide a sample large enough to identify trends, patterns about the research proposed and actually carried out as well as other structural items.[8] The first step was to identify distinctive expectations, against which potential achievements could be measured. The ERC had been founded with a range of objectives, at least if the most authoritative legal text were to be taken seriously in this regard: to 'reinforce excellence, dynamism, creativity', to 'improve the attractiveness of Europe for the best researchers', to 'put European research in a leading position', to create 'new [. . .] scientific and technological results', to 'stimulate the flow of ideas', and to 'allow Europe better to exploit its research assets and foster innovation'.[9] In 2009, the Secretariat came up with an attempt at operationalizing them into a coherent 'monitoring and evaluation strategy', with the intention 'to generate a broad and integrated understanding of ERC's performance and impact'. The strategy distinguished four dimensions, allegedly 'correspond[ing] to four objectives of the "Ideas Programme"' (see, however, Figure 7.1).[10] Monitoring 'performance in science management and organization efficiency' was a simple way to reuse data that had to be produced anyway, since the ERCEA, as the administrative branch within the ERC compound, was obliged to report annually to the European Commission on the management and conduct of the programme it was destined to operate.[11]

<table>
<tr><th>Objectives according to Ideas Programme[12]</th><th>'Four dimensions' of ERC 'Monitoring and Evaluation Strategy'[13]</th></tr>
<tr><td></td><td>(1) performance in Science Management and organization efficiency</td></tr>
<tr><td>Open the way to create 'new [. . .] scientific and technological results and new areas for research'</td><td rowspan="3">(2) advancing frontier of knowledge and training</td></tr>
<tr><td>'stimulate the flow of ideas'</td></tr>
<tr><td>'reinforce excellence, dynamism, creativity'</td></tr>
<tr><td>'improve the attractiveness of Europe for the best researchers'</td><td rowspan="2">(3) impact on researchers, research institutions and policies</td></tr>
<tr><td>'put European research in a leading position'</td></tr>
<tr><td>'allow Europe better to exploit its research assets and foster innovation'</td><td>(4) socio-economic impact</td></tr>
</table>

Figure 7.1 Correspondence of 'objectives' to operational 'dimensions'

Compiled by the author; see endnotes for references.

As for the three other dimensions, their formulation was driven by an attempt to square what would be feasible from the ERC's own data with what would produce solid evidence of the ERC fulfilling its objectives.

The second dimension, on 'advancing frontier of knowledge and training', was easy to achieve. The fact that academic publishing had become somewhat standardized (at least in the domains of the hard sciences as well as in parts of the social sciences) made it possible for the ERC Secretariat to look for certain patterns within established databases such as the Thomson Reuters Citation Index or Elsevier's Scopus. Following the NSF's 'index of highly cited articles', the respective analysis showed that '13% of ERC publications from 2008 to 2010 were in the top 1% most cited publications in 2012'.[14] Though still preliminary results, they signalled that the ERC had fulfilled its ambition, namely, to reach top-notch scientists across Europe.[15]

As satisfying as those findings were, they could not be used to claim an 'impact on researchers, research institutions and policies' or a socio-economic impact even, with which the latter two dimensions were dealing. As was well known to the Scientific Council and the ERCEA management, beyond descriptive statistics on budget distribution, profiles of applicants, and bibliometric analysis, systematic evaluation of what had been funded and its role for further scientific progress would be difficult. To tackle the problem, the Scientific Council had commissioned in its earliest Work Programmes projects 'to analyse the impact of the ERC, analyse the functioning and performance of the ERC and to assist in the development of a strategy for the monitoring and assessment of ERC activities'. The hope was to establish 'an appropriate monitoring, assessment and evaluation framework' based on the 'independent exploratory work' of academic scholars.[16]

Four projects were eventually selected: two represented scientometric avenues, reflecting a genuine concern regarding content and impact: most notably, was the ERC really funding research at the frontier of science? Was it detecting the 'right' proposals? One project was based on 'the assumption that excellent truly creative research can be found in particular in these emerging areas', and inquired 'to what extent the activities supported by the ERC cover and contribute to these research areas'.[17] Another was to develop 'a scientometric-statistical model [. . .] for inferring attributes of "frontier research" in peer-reviewed research proposals' submitted to the ERC.[18] Both these projects were methodologically rigorous, and they also produced some academic papers.[19] Yet their results were hardly convincing, and initial hopes of the Scientific Council and of the Secretariat, particularly concerning new avenues for measuring the accuracy of the ERC's funding mechanisms and the impact of the ERC were not met. The problem was not so much data in itself – for the material of proposed aspects was available – but restricted access due to legal constraints and methodological complications due to the material's nature.[20] Even more substantial was the statement that there were 'clear limits to the use of bibliometrics to measure frontier research and emerging research areas'.[21]

Another route was to investigate the impact of ERC funding on researchers and research institutions. The two remaining studies were commissioned to provide answers to that; the first set out 'to develop a novel methodology for the study of the impact of research funding schemes on knowledge and its social conditions;'[22] it provided some generally useful insights into the issue albeit, due to its exploratory character, its results remained to be validated. The second study was a long-term survey on the effect that the 'Starting Grant' scheme had on its recipients. Focusing 'on the individual perspective' and painting 'a broad picture concerning the questions of whether the [Starting Grant] programme succeeds in attracting up-and-coming 'excellent' young researchers from all over the world', the study found, among other things, that applicants in the early-stage career track call 'already exhibit an above-average output prior to the application', that the funding 'is seldom used to enable mobility', and that it 'has strong positive effects on the individual career development of postdocs'.[23]

Ultimately, the approach to evaluating the effects of the ERC yielded robust data that, through academic publications and the reports, would feed indirectly into the general reception of the ERC.[24] But was there evidence of impact? Even though a policy letter came to the conclusion that 'early evidence suggests wide-ranging effects on the science system' in Europe,[25] another publication from the same project team was more reticent on that question, because, in some respects, it was still too early to say:

> ERC funding shares a feature with most institutional innovations of the last three decades. It is flexible in some dimensions but a 'one size fits all' – institution in others. It is build [sic!] after a blueprint derived from the biosciences or, more general, smallscale collaborative experimental science. It thereby adds a new standard of funding that is deemed applicable to all disciplines. The effects of this new standard are yet unclear.[26]

In addition, the studies provided little new support to the methodological quest of how to continuously measure impact of a funding instrument such as the ERC; as one of the reports openly acknowledged,

> it will probably be difficult to assess the work of the ERC by assessing the extent to which impact has really occurred (at least for some time). It may be more appropriate to develop evaluation techniques focusing, at least initially, on the conditions for impact to occur.[27]

Despite those setbacks, the Scientific Council pushed on with appropriately processing the stream of information about its funding activities for the public. In 2012, it set up another working group, this one specifically dedicated to 'Key Performance Indicators'. Its chair was Reinhilde Veugelers, a fresh member of the Scientific Council, and an expert in the field of innovation policy.[28] In the sub-group's first gathering, she explained the purpose of such indicators for the ERC:

> justifying to stakeholders its resources; improving the internal governance of the ERC; improving the governance of the EU research funding system.[29]

Establishing the sub-group was, again, a reaction on political pressure to prove the ERC's contribution to innovation policy: with the upcoming edition of the Framework Programme, each of the European Commission's funding programme would be required to report 'key performance indicators' to the parental DG.[30] Even though the evidence for impact was not (yet) to be found, the Scientific Council was confident that there would be plenty of material to engage in the evolving discourse about innovation policy at European level and to emphasize the ERC's unique role in it.

7.2 i-conomy

Ever since the adoption of the Lisbon Strategy, 'innovation' had remained high on the priority list of EU politicians. From 2010 on, another wave of innovation-related policy papers was produced.[31] This was not coincidental, as by then, the initial Lisbon Strategy had come to an end, and the European Union was in need of a new policy framework to refer to. Despite weak results, there was broad political agreement to continue the promises of growth through innovation that had characterized the Lisbon accord. Partly, the decision was supported by a morale-boosting argument, namely that '[u]ntil the crisis hit, Europe was moving in the right direction'.[32]

The crisis, of course, referred to the implosion of the US housing market and its severe implications for the European banking sector (which, it should be noted in passing, had greatly increased its exposure thanks to, what was called, 'financial innovation'). Even the alleged progress was not undisputed; several indications had been raised – and already earlier than 2008 – that major areas would not be able to come up with the ambitious goals of the Lisbon Strategy.[33] Most importantly, the belief in innovation was unabated. A study by economists close to the Commission was quick to point out that '[s]topping European research programs would cost each year at least 0.7 % of GDP and 380,000 jobs from 2025 onwards'.[34]

The new initiative was labelled the 'Europe 2020 Strategy', again shooting for ten years. It aimed at 'smart, sustainable, inclusive growth' and was composed of several 'flagship initiatives', of which one was meaningfully called 'Innovation Union'.[35] Much of the fuss around innovation during this time would probably have to be ascribed to political rhetoric, with the second Barroso College of Commissioners just at the beginning of its term attempting a head-start. How much heightened expectation there was around the topic could be seen from a rather unsuccessful quip used by Geoghegan-Quinn around that time; the Commissioner spoke of Europe becoming an 'i-conomy', meaning 'a really vibrant innovation economy'.[36] She soon dropped it from her talking points, but it nonetheless nicely mustered the promises that were by then expelled by policy-makers across Europe.

There was, however, an organizational effort behind the topic, too. The 'Europe 2020' strategy decreed that, within the Innovation Union flagship, '[e]very link should be strengthened in the innovation chain, from "blue sky" research to commercialization'; the ambition was '[t]o improve framework conditions and access to finance for research and innovation so as to strengthen the innovation chain and boost levels of investment throughout the Union.'[37] The Commission was following up by proposing, through open consultation, a 'Common Strategic Framework' of the 'various programmes' that 'support research and innovation, covering activities across the innovation cycle [sic!]'. The goal, ultimately, was 'a simpler and more efficient structure and a streamlined set of funding instruments covering the full innovation chain [sic!] in a seamless manner.'[38]

From a purely scholarly point of view, it hardly mattered whether innovation was depicted as a 'chain' or a 'cycle' – both were just contemporary metaphors for what was long known among scholars concerned with the elusive phenomenon of innovation as the 'linear model'. That model acknowledged that technological advancement was somehow dependent upon the production of new knowledge, which, through many interacting social and circumstantial variables, would eventually contribute to economic growth. The model had long been criticized for being an image too simple to simulate how new knowledge was transferred to commercial products and increased productivity. Still, the 'linear model' had also become a 'social fact', affording 'administrators and agencies a sense of orientation when it comes to thinking about allocation of funding'.[39]

When the ERC Scientific Council leadership deemed it necessary to intervene in the on-going debate on innovation in order to make a clear statement for its own cause, the metaphor of a cycle was crucial. It allowed them to claim that while the distinction between 'the production of scientific knowledge and innovation' remained valid, there was 'a long but never linear trajectory linking the two'.[40] The statement then admitted a core problem of funding instruments for academic research generally, namely that 'public expectations of short-term results' could often not be fulfilled by (what the ERC prominently expected to fund) 'frontier research'. To make good of it, the statement introduced two types of innovation. The first was 'incremental' innovation, meaning a 'process of improvement of already existing products and components'. The other was called 'radical', and it entailed 'paradigm changes in the ways in which societies and their economies are organized and grow'.[41]

By the time of this contribution, the Scientific Council had taken an important step to reinforce the self-image of being an enabler of radical innovation. Early on, it had established a working group on 'Relations with Industry', under the leadership of Jens Rostrup-Nielsen; this sub-group then suggested making a (rather symbolic) contribution from the overall ERC budget to the innovation efforts by setting up an 'add-on' funding line; the so-called 'Proof of Concept' grant was implemented in 2012 exclusively for ERC grantees with the aim of 'covering the funding gap which can occur at the earliest stages of an innovation'.[42]

In the ensuing discussion around the Innovation Union, this funding line came in handy. Yet it would be necessary to detail what was meant by 'radical innovation'. In a contribution to a pamphlet on the very topic, Helga Nowotny called it 'a discontinuous, deeply disruptive form of innovation', one that was 'almost entirely due to new scientific insights, discoveries and technologies made in basic research'. And once found, this kind of innovation 'has enormous repercussions for the structure of our economies and their growth potential' – the classical example being the laser.[43] As Nowotny admitted,

> we cannot draw a concrete link between specific scientific insights and increases in productivity. But there is no doubt that investment in basic research contributes significantly to the processes that lead to innovation and productivity.[44]

Accordingly, and also slightly more self-assuredly, the Scientific Council's report asserted that 'practically all epoch-changing innovations are the outcome of major scientific and technological breakthroughs.'[45] Thus, it envisioned a concept where commercialization could be achieved through new insights based on scientific evidence.

In the end, the ERC's intervention served one simple purpose: to keep the ERC in the political discourse and to claim its stakes in the negotiations about its place in 'Horizon 2020', as the eighth edition of the Framework Programme would be called.[46] Since it would indeed represent a more inclusive approach, at least from the way it was structured, the ERC would have to make sure to be represented appropriately; the way this was done – the ERC would come to rest in the so-called 'excellence' pillar within this Framework Programme's edition[47] – was also to acknowledge the aura that the ERC had created.[48]

There were also more offensive measures taking place to safeguard the funding instrument's position within the diversified funding landscape of the EU. In that respect, it was crucially important to maintain good relationship with industry, in order to prevent criticism from that side about public funds wasted (that is, spent for academic research instead of for technological development). Then later, the ERC leadership would embark on a cooperation with the World Economic Forum (WEF), offering the opportunity to present the ERC to industrialists and business people from across the world.[49] Similarly important, however, was to maintain good relationship to academic circles and scientific institutions, and to exploit the ERC's standing among researchers across Europe.

When the preparation for the eighth edition of the Framework Programme began in earnest, and after signs of member states wanting to reduce the overall commitment, an (allegedly grass-roots) campaign was organized in order to secure a good budget in the intricate negotiations for the next MFF. The ensuing media campaign included a letter signed by Nobel Laureates and more than 150,000 signatures to a petition demanding 'no cuts on research'.[50] This time formally led by Wolfgang Eppenschwandtner, ISE had once again provided the institutional carrier; however, the campaign was coordinated by Nowotny and covertly supported by the communication unit of ERCEA. The ERC's overall budget within the next edition of the Framework Programme was not increased as much as many would have hoped for, but it provided enough for the ERC to keep its existing funding tracks predictably stable over the next seven years.[51]

In the 2013 version of its annual report to the European Parliament, when the ERCEA reported its 'findings', the report began by saying that 'eight Nobel laureates and three Fields Medalists [are] among its grant holders', and 'a total of 134 ERC grantees have received other prestigious international scientific prizes and awards'. Furthermore, the ERC could claim that, based on the large number of projects that it was funding by now, more than 3,000 ERC-funded researchers produced new knowledge, with a promise to provide for more innovation (whatever was meant by it): 'over 20,000 articles acknowledging ERC-funding have appeared in peer-reviewed high-impact journals between 2008 and 2013'. And 'each ERC grantee employs on average six other researchers, contributing in this way to the training of a new generation of excellent researchers'. In addition, it was also deemed important to note that 'around half of all ERC team members hold a nationality that is different from that of the Principal Investigator of their project.'[52] The selection showed what the ERC expected policy-makers to think would be significant indicators for fulfilling its objectives; and it also indicated the achieved routine to confirm its own position.

7.3 Symbolic value

Increasing circumstantial evidence showed the ERC to be a relevant funding instrument for researchers, research organizations and policy-makers. Thus, when the routine evaluation of the seventh edition of the Framework Programme was taking place, a 'support paper' specifically dedicated to the ERC detailed the achievements and came to the conclusion that it had been 'highly successful'. Not surprisingly (given the fact that the report had been written by an economist specializing in analysis of science and technology), the text drew on the argument of Keith Pavitt and argued that Europe 'has long been in need of this institution' in order to create 'the intensity of competition experienced in the USA'.[53] Even though the overall report of the High-level Expert Group on assessing the seventh edition of the Framework Programme was critical in some nuances, it also lauded the ERC as a 'success'.[54] In its reaction, the Scientific Council still expressed its fear that, in the report, the 'unique objectives and features of the ERC [. . .] tend to be diluted if not obscured'.[55]

The relevance could be distinguished along three lines. First, the instrument was becoming a major force for influencing the academic career of early-stage researchers within the EU; or, to be more precise, in those parts of the EU to where the ERC money was actually flowing. Not least, the long-term evaluation on the impact of ERC funding in the 'Starting Grant' scheme, as well as the enthusiastic reception on calls under this scheme, had shown that funding in the early-stage career track provided grantees 'with the opportunity to complement [their high level of cognitive independence, TK] with independence with regard to the allocation of material and human resources as well as laboratory and office space.' In other words, they were freer to negotiate their place in the hierarchy of their host institution, and more often than not, they were taking up this opportunity, too.[56]

The relevance of the funding track for early-stage career scientists was also reflected by the development of the ERC's internal annual budget allocation. While, in the beginning, the funding for the senior scientists' track was almost double, it would be the other way round in the years 2013–14 (see Figure 7.2). The ERC leadership reacted to the increasing demand by directing more money to the juniors: over the first period of its existence, the ERC had funded almost 3,400 projects in the early-stage career track, while a little less than 1,900 in the senior track. Interestingly, and despite its efforts to go global, the ERC had remained a European-centric instrument.[57]

A second area where the ERC was acknowledged for its relevance concerned the fact that it helped raising expectations about the quality of scientific research. Probably the most important effect here was that the ERC 'reinforced the importance of internationalization in peer review, and overall, reinforced cross-European competition.'[58] To do so, the ERC Scientific Council had skilfully achieved the establishment of a certain kind of review culture, not least because of the careful selection of the panel members and the panel chairs. This strategy helped the ERC to position it as the unique selling point in the crowded landscape of European funding opportunities, with the ambition 'to galvanize Europe towards a merit-based competition for the most talented ideas and researchers';[59] and to 'induce national councils to develop themselves in similar directions.'[60] Whether this culture really was diffusing into other organizations remained to be seen.

The ERC turned also out to be a reliable 'innovator' (if that word

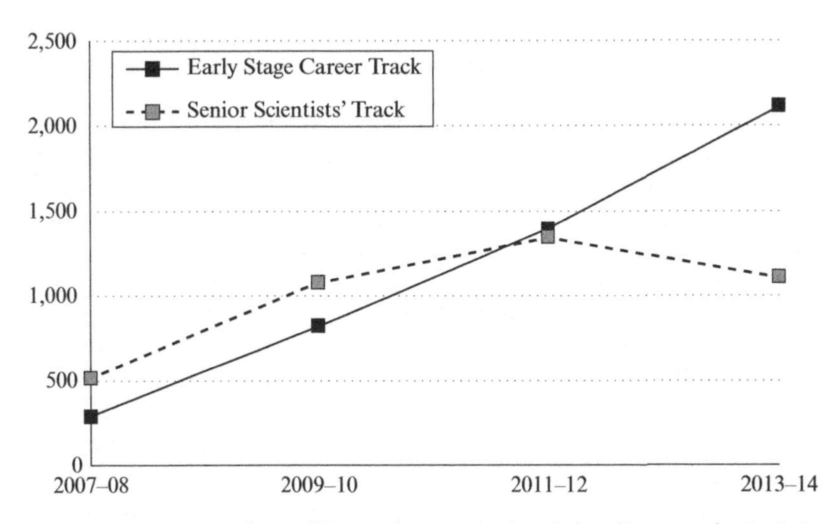

Figure 7.2 Budget allocated to main ERC funding tracks in M

Compiled by the author; data from ERC Work Programmes and Annual Activity Reports; Early Stage Career Track combines budget allocation to the Starting Grant calls and the Consolidator Grant calls; Senior Scientists' Track consists of budget allocated to the Advanced Grant calls.

can be used in this context) in relation to the rest of the FP format's funding programmes. Specifically, it 'proved to be user-friendly' and turned out to have 'fast granting and payment procedures'.[61] In the preparation of the Horizon 2020 programme, the Commission took up the previous edition of the crucial innovations implemented by the Scientific Council for the ERC already under FP format, namely a single reimbursement rate and a flat rate for indirect costs.[62] Finally, the ERC had become a relevant player in that the projects that it funded were increasingly used for European-wide benchmarking exercises, as 'the symbolic value of the ERC is already vast for European research organizations'.[63] The ERC grants became a measure of relative success or failure across Europe by simply comparing the number of grants awarded to a nation, or, even more significantly, between individual host institutions. For example, a study of indicators for research excellence in Europe resorted to four variables, three of which were long-established practice; the fourth was based on the ERC:

> the number of highly cited publications published by a country, the number of high-quality (i.e. PCT) patents on which a country is listed,

> the number of world class universities and research institutes in a country, and the number of ERC grants received by a country.[64]

The findings of any of those benchmark exercises were not exactly comforting the Scientific Council, as they showed 'a strong concentration of beneficiaries'.[65] How could it be justified that 25 per cent of all ERC grants were located at host institutions in the United Kingdom, a country holding not even 13 per cent of the EU population, while, at the same time, the thirteen states that had accessed the Union since 2004, representing more than 20 per cent of all EU citizens, only held a share of slightly more than two per cent? The answer was to stylize the ERC as merely the messenger of a deeper-seated problem. As Fotis Kafatos and co-authors acknowledged already in an early analysis, the ERC's 'grant distribution reflects the reality of unevenly distributed national R&D investments across Europe'.[66]

To link the distribution of ERC grants across European nation-states with the country's respective R&D funding level highlighted that, only with appropriate funding from home, success in the competition for ERC funding could be expected.[67] That was certainly a valid argument and also a framing that, overall, was broadly accepted. The reviewers of the seventh edition of the FP format, too, came to the conclusion that

> the correlation index between the number of [ERC, TK] grants per country and the Global Effort in Research and Development (GERD) [sic] of the country is 0,81, which indicates a strong correlation. There was also a very strong correlation of 0.97 between the number of grants per country and the number of publications within the 10% most cited.[68]

Yet, while correlating ERC fund distribution with statistical aggregates showed that countries performed according to what was to be expected, it could also reveal some further imbalances between the performance of individual countries (see Figure 7.3). Theoretically, those imbalances could be blamed on undetected biases in the decision-making procedure. Since the latter was widely regarded as robust and fair, those imbalances would typically be explained as specific weaknesses or strengths of the respective national innovation systems. Still, the ERC could not completely dilute the fact that it was not only depicting the structural imbalance of resources available for research in different parts of Europe a certain constellation, but also reinforcing it through its own, highly valuable, funding. Playing with the role of the messenger would only be successful if the ERC

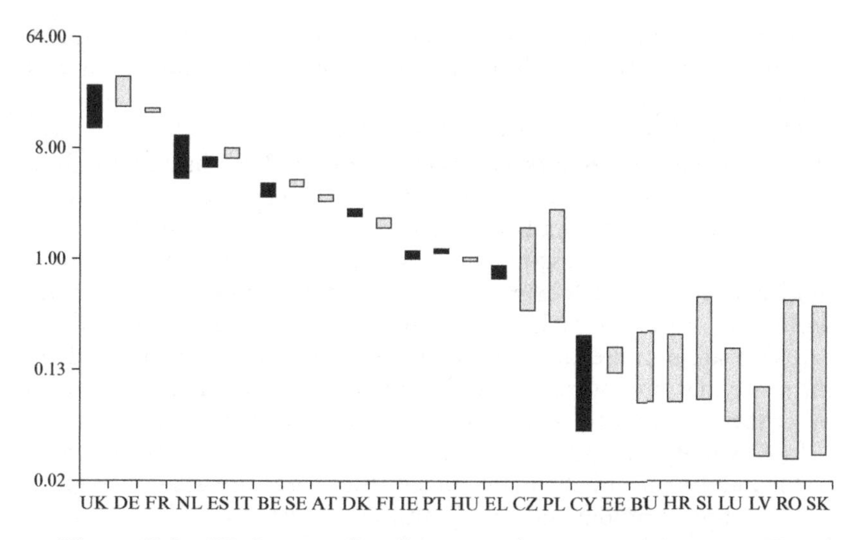

Figure 7.3 ERC grant distribution to host institutions per EU member state correlated to GERD

Compiled by the author; data: ERC statistics, 2007–2014, and Eurostat (on Gross Domestic Expenditure on Research and Development [GERD], based on Purchasing Power Standard); dark grey bar: share of ERC grants higher than expected; light grey bar: share of ERC grants lower than expected; vertical axis in logarithmic scale.

would not only provide data for external evaluation, but continue to interpret it, too.

In 2015, the ERCEA published two colourful reports, one of which was to depict 'key facts, patterns and trends', while the other promised to analyse the research that had been funded thus far.[69] The first report was little more than a dry compilation of descriptive statistics on 'demographic profile of applicants' (with the exception of one funding call, 'women are less successful than men in obtaining ERC grants'), 'project costs breakdown' (personnel costs 'range on average from 50 to 60% of total project costs'), 'time-to-grant' (on average being '363 days').[70] It also showed the 'success rate' of the ERC funding calls, pointing towards some 'variation' 'across different scientific disciplines as mapped by the scientific panels assigned to the evaluated proposals'.[71] Those were fascinating facts for everyone involved in the operations of the ERC; for anyone outside, the numbers confirmed the message that, by now, the ERC was a well-oiled machine running thousands of project proposals through its evaluation process by year, that it was spot-on the potentially detrimental effects of its decision-making procedure, and that it was extremely competitive.

The second report promised to reveal the 'science behind the projects' that the ERC had been funding between 2007 and 2013. To a large extent, it was following an ERC-centred perspective, looking at which academic fields had been funded in which ERC panel slot, and highlighting some 'cross-panel/cross-domain interactions', meaning funded projects that 'cover areas of research beyond those that are within the remit of each individual panel.'[72] As such, the findings revealed some interesting topics, but remained statistical artefacts in the sense that they were not applicable for comparative analysis with other funding instruments. Still, the report made some nice press.[73]

Interestingly the report also showed how 'the ERC contributes to particular European thematic policy priorities'; the examples given were 'nanotechnology, energy, health and migration.' That was not by chance: all four 'are high priorities in the 'Industrial Leadership' or the 'Societal Challenges' pillar', as the report maintained, referring to other programme lines under the eighth edition of the Framework Programme. Accordingly, it showed 'how a wide range of ERC panels are contributing to address the broad challenges, sometimes even across the whole spectrum of research domains.'[74] The reference was remarkable, as those priorities were established through the very political negotiations that characterized the rest of the FP format and that the bottom-up approach of the ERC had actually intended to avoid, simply because they were not based on scientific reasoning. It revealed to what extent the ERC was willing to conform with the expectation to contribute to the overall political ambitions of tackling those 'grand societal challenges', even though it was exempted from it in the legal text.[75]

— 8 —

SUMMARY

The aim in writing this book has been to provide a more nuanced picture to the often-heard slogan of the ERC's being a 'success story'. As the book suggests, the key to this success lies in the specific answers to three core questions in relation to the aura that the ERC advocates, together with what the DG Research leadership of that time had sparked. The instrument created as a result of this campaign was destined to fund 'frontier research' based on the sole criterion of 'excellence'. The three questions concerned: first, how the aura's promise emerged, how the ERC idea was argued and brought to political relevance, and how much this depended on coincidence and context; second, how the demands of appropriately institutionalizing this aura questioned existing organizational frameworks and demanded new solutions and, even more importantly, the will to compromise; and third, how this aura was kept, and actually reinforced by way of routinizing a robust and legitimate procedure of decision-making for distributing the ERC's funds.

The ERC campaign was an extraordinary effort to push the idea of an independent funding instrument at European level onto the political stage. It required the context of a unique phase of heightened expectations in research and science for the further progress of the European integration, so that the ERC idea could eventually strike roots. The ERC advocates, top-notch academic jet-setters most of them, represented a distinct milieu within the European elite, and they felt that, until then, the European project had missed out on them. As such, they held naive ideas about Europe alongside realistic assumptions about how to get what they wanted; but the advocates alone would not have had the political skills as well as persistence to mould the ERC into the European institutional landscape. It could only be brought through those procedures due to the fact that, after a period of reluctance, the Commission had embarked on the campaign.

The price of that partnership was that the ERC would end up firmly under the wings of the Commission; much more firmly than the advocates had expected and hoped. Exposing the 'radical proposal' to the Union's political cycles and decision-making procedures inevitably pulled the ERC into the orbit of the FP format; yet by playing this game skilfully, the Commission also erected a new body that, once established, was determined to secure its autonomy – the Scientific Council. While the conflict between the advocates and the Commission team around Achilleas Mitsos was eventually settled, it was an indicator that the tensions would be far from over.

The Commission joining the campaign is the irony that marks the pre-history of the ERC: having been an instrument imagined by scientists for scientists, the ERC made it through the political realm only because of the stewardship of the European Commission. There was no other option, really, as the only two other approaches – either to anchor the ERC in the European constitution, or to establish an ERC as a multi-national effort – were doomed. Then, the campaign was ultimately absorbed, and successfully integrated, into the institutional framework of the European Commission, while continuing to insist, and with some success, that it was different still. This difference within the same was carved out through intricate negotiations on several layers, not only during the initial phase of implementing the ERC, but also later, and even today, the process is probably still going on.

The answer to the second question was about resolving expectations about what the remit of the different entities within the ERC compound should be. At its centre was the Scientific Council, primarily because of its unique status – although nobody really knew where its remits and limits were, this newly founded body, composed of highly distinguished scientists and scholars, fully embraced the ERC idea and perceived it as a huge opportunity to leave a mark on European R&D funding policy. Over the first fifteen months, an important achievement of the group was to find a core set of common convictions, despite various arguments internally; another achievement was to set up the informal rules of governance within the emerging ERC compound before the political machine of the EU could adopt the legal texts.

The ERC governance structure had been intended by advocates to be autonomous and independent, but it turned out to be a compound of legal entities based on different principles. That solution was an adequate compromise for the European transnational polity reality, but it could not be established without frictions. Even the so-called Mid-Term Report turned out to be a mixed blessing in that respect. It would force the Commission to lay bare its operations; at the same time, the prospect of the review would continue to keep key actors concerned – and, inevitably, initial doubts would be reignited. In the long run, however, the autonomy equation underlying the ERC governance turned out to be reliable; specifically, it turned out that, what the Mid-Term Report had denounced 'old-fashioned division of labour' between the Scientific Council and the Executive Agency, as 'sub-optimal' and 'an obsolete model of management which makes a sharp distinction between decision-making and execution' was actually working pretty well.[1] This was due to Fotis Kafatos' legacy, albeit in some unforeseen ways.[2]

If the ERC's pre-history was marked by an ironic twist at its core, arriving at a compromise with regards to the governance structure was more of a tragedy for some of its protagonists. While the career and notoriety of Robert Jan Smits and Helga Nowotny, respectively, was significantly boosted by their engagement with the ERC, others have been paying a high toll. William Cannell, ambitious and having fully incarnated the spirit of the ERC, was stripped of many responsibilities when Directorate S was set up; he left the Commission a couple of years later. Ernst-Ludwig Winnacker was excluded from the ERC's main management meetings, and, when his term was coming to an end, he would not even have been financially compensated. Winnacker moved soon to another appointment; but his deep frustration was later summed up and published in a personal rant.[3] Fotis Kafatos had relentlessly aimed for compromise with the Commission, insisting nonetheless 'on our decision-making responsibility', as he once put it.[4] Being in frail health, he had occasionally wavered;[5] in early 2010, he decided to step down from his position.

The ERC was more than the result of a skilfully conducted campaign motivated by dry calculations; in the European contest, it was ideologically something completely new. So far, the common understanding of R&D policy had restricted the European Commission from distributing funds to 'pre-competitive research' focused on socio-economic impact (i.e. innovation), and with the implicit promise of fair shares going back to each member state. The absence of fair returns had been the political weak spot of the Commission proposal. But at the very time that the Council of Ministers reluctantly accepted the new mechanism, it turned into the ERC's most powerful feature:

scientific excellence. The third part of this book investigated how this excellence was operationalized in the ERC's 'scientific strategy'.

In his opening speech to the Scientific Council, Commissioner Potočnik pointedly had called it the 'most important' task [. . .] to 'establish[] the credibility of this Council in the eyes of the scientific community'.[6] Credibility would indeed be crucial for a policy instrument, which had been devised to disburse large amounts of public funds to rather obscure projects of scientific research. To meet the goal, it would be necessary to get strategy and operations right; but that would not have been sufficient. In parallel, the ERC also required marketing – transferring the mere promise of its aura into a commonly shared perception of a successful policy instrument. That was achieved primarily through a very sophisticated, i.e. resource-intensive, constantly fine-tuned evaluation procedure, which was diligently orchestrated by the Scientific Council; and by skilful interventions into the on-going discourse of European innovation policy that brought forward convincing arguments emphasizing its importance and necessity.

Key to the ERC's efforts was the notion of excellence; crafting its meaning – in terms of strategy setting – remained one of the Scientific Council's most important responsibilities. The eighth edition of the Framework Programme unified the ERC with other funding instruments under the common label 'excellence'.[7] It was a signal of the Commission's overall appreciation for the ERC; and, since the ERC constituted more than double the amount allotted to other programmes within this new heading, there was little doubt that it would continue to lead the efforts of shaping its meaning. Yet marking the ERC's uniqueness came with a price, too. With the ERC blasting billions of euros to fund research across the European Union, the third trend of this book is about what could be called the routinization of aura.

The internal discussions of the Scientific Council revealed some inherent weaknesses in the specific peer review procedure as implemented by the ERC. For example, it was vexed by the difficulty in creating a common understanding of the complex concepts that were put in place, whether it was called 'interdisciplinarity' or 'synergy'. As a reaction, the Scientific Council tended to take a more critical stance towards broadening its funding portfolio, claiming that simplicity would be the best way forward. Similarly, when pressed to prove its intended impact as a policy instrument, the ERC would, by its own choice, occasionally conform with scientific challenges that were politically imposed.

The introductory chapter claimed that, as a comprehensive story of the ERC, this book would also be a case study on the affairs of research policy in Europe, and the relationship between science and policy more generally. Is the ERC a success *of* European integration, or one *despite* it? Of course the ERC is a direct result of integration, for, obviously, if there were no politically powerful concept of a transnational space such as 'Europe', there would not have been the faintest idea of a European instrument for funding research.

The matter becomes more complicated when one looks not only at the concept, but adds its institutional reality, i.e. the European Union. In its volatile political context, the advocates started campaigning for the ERC, and despite many odds and little prospect of realization, they had an infallible instinct for the right timing – and a punch line, namely, their common disgust at the way research funding was conducted in Europe thus far. Lamenting the EU's inflexibility, bureaucratization, and political meddling in scientific affairs was a clever way of gathering support (the proximity to anti-EU campaigns nowadays rocking national arenas across Europe should be mentioned, with the notable difference that the ultimate goal was to achieve a better Europe and not less). To some degree, thus, the ERC had initially been based on a populist impulse against existing funding policies.

The ERC is also a case study on an even deeper-seated arena of negotiations, the intricate relationship between science and policy. The story laid out here ultimately refers to the ERC being an instrument to make use of science for the greater cause, or being a channel to pump money into labs and study rooms. To begin with, the ERC was somewhat a counterintuitive initiative in a period when, more than ever, politicians wanted to spend money for targeted, large-scale undertakings such as the genome project, the war on cancer, the fusion energy project, and so on; and it was also outfitted with a degree of liberty that other funders (such as the NSF) saw scrutinized during that very same period, namely to fund research based on excellence only.

Yet there was a mundane reason for this: while US science policy (and, subsequently, that of most other industrialized countries) had evolved from funding 'basic' (academic) research towards more relevance-oriented priorities, it was the other way round in the transnational European polity: the ERC had been justified as filling an important gap in the existing portfolio of funding opportunities, namely to address academic ('frontier') research. And since, other than the NSF, the ERC was part of a larger research-funding instrument, it could be more easily protected within the Framework Programme format to play its specific role; its philosophy of simplicity is also an expression of this. However, why is simplicity such a tempting attribute? Ultimately, it seems as if the reason is to be found in the European integration process itself, where '[s]cale in the face of diversity has produced a complex – almost incomprehensible – polity'.[8] This impulse also has some hindrance on the development of the ERC itself. The best example has been the on-going discussion about new funding schemes in the Scientific Council. Caught in the FP format and determined to stay true to its initial niche, the ERC cannot become a fully fledged funding agency also because it is narrowed down to a very specific set of funding by its very own strategy-devising body.

Maybe after the aggrandized rhetoric of the early years, the ERC's niche within the Framework Programme is fitting. This all the more so since the European integration project has moved on and now entered a period distinctively darker than before. In the European Union, talks of crisis had always abounded. But the fiscal problems, rising unemployment and migration movements; the rise of populist anti-EU parties in many countries, the vote against mass immigration in Switzerland and the prospect of the UK leaving the Union; the economic gap between Germany and other member states and the financial meltdown of Greece; the war in Ukraine, the turmoil in the Middle East leading to the Syrian civil war, and terrorist attacks in European city centres; all those events have had their role in the fact that innovation policy, which had been such an important and seemingly urgent political feature of the EU before, somewhat lost its zeal recently.

On the other hand, the Union's policy instruments were often used as an argument to speak for the continuation of the integration project. Maybe most significantly, the ERC had gained a tiny yet relevant role in the political frictions that were increasingly shaping (and rocking) the European integration project in general. After the surprising majority of the Swiss electorate to adopt the 'initiative against mass immigration', and due to the restrictions imposed by the Swiss, the European Commission decided that Switzerland could no longer participate as an equal partner in the eighth edition of the Framework Programme. Since researchers based in the country had been particularly successful in the competition for ERC grants, and since Swiss universities had begun to rely on the influx of ERC money, the academic institutions in this small country were frantically urging their government to negotiate a way to get access to the ERC funding pot again. Similarly, among academics and research managers, the run-up to the crucial vote about the UK leaving the EU was greatly debated in relation to the potential impact that this may have for the funding level of British universities.[9] In a way the ERC has raised the EU's purchasing power – at least among academic elites.

The ERC history is, ultimately, also a study of the different realms of negotiation between scientists and policy-makers in realizing the distribution of significant funds for academic research. It is, thus, an example of how the seemingly timeless concepts of autonomy and accountability, excellence and relevance, serendipity and usefulness, are interpreted and operationalized in the early twenty-first century. Much of this book, thus, has been about the political discussions around those concepts, and I believe it provides much material for further investigation in that direction. Yet I want to end with a personal observation that also rectifies, to some extent, the autonomous realm of scientific deliberation within the bureaucratic environment in which it is settled and which attempts to control everything.

— 9 —

POSTSCRIPT

As mentioned in the Introduction, I have been in the extraordinary position of being able to observe, in a non-participatory role and without further obligation, numerous meetings of ERC evaluation panels in the ERCEA headquarters in Brussels. From the beginning I have been fascinated by the many dimensions of peer review in research funding that I experienced. It is, after all, the central scientific activity around which all other aspects of a funding instrument, such as the ERC, is built. After having analysed at great length the politics of research funding in different perspectives, I want to end this book with a personal note on this core activity.

The extent to which the peer review procedure today requires heavy machinery of coordination and constant orchestration is breathtaking. Even a single panel meeting depends on so many prerequisites, and depends on so many different rules specifically designed for this kind of gathering; it follows a clear and easily definable goal, and yet it is a temporal space where the unexpected is happening. There is a play-off where rules, formally written ones and, even more importantly, informal ones, are interpreted, arranged, spontaneously formulated, judged and tossed away. For those involved (most notably, the panel members), each meeting remains a highly emotional affair, closely related to their conviction to act in the greater interest of science.

This, and almost everything else on the panel's work of allocating funds has already been accurately analysed in the respective studies by Michèle Lamont and her collaborators, who found, most importantly, that panels, if composed of the right persons, are primarily characterized by a shared belief in fairness.[1] The one notable observation that I would like to add to this body of work is that the established set of different informal rules for everyone present in the room forms a powerful magnetic force.

Initially, when I started my observations, I could not believe how dull the panel meetings were for most of the time. Yet I soon realized that all the boredom, confusion and awkwardness may have been necessary to lead up to the decisive moment. That moment is when one of the panel members questions the basic assumptions of those present. The question is always different – what is our understanding of excellence? What do we mean by outstanding? What do we look for in a proposal? But its intention is always the same. Still, the need to raise this question (and to raise it again, and again, in every new panel meeting) seemed strange and out of place to me. What, if not to have ready-made answers to that particular issue, are all the experts in the room supposedly experts of?

Eventually, I came to learn that the question is sort of a quality stamp, a sign that the panel is on course to achieve its goal. To reach the moment when someone in the room asks the question was decisive, by which I mean that it is all the panel meeting is about: it is basically built towards this unique moment. Whenever the question was asked, an answer was eventually found. The problem, I realized, was when the question was not asked, which happened occasionally. It meant that either the panel members were too detached from the proposals, or the group was overwhelmed by too many proposals and deferred to a mechanical approach of working off.

If the question is the decisive moment, it is the discussion evolving from it that constitutes the sense of community between the panel members. Maybe that's because the question – whatever its actual direction – is a fundamental one; yet it can only be answered by concrete examples, that is, the project proposals to be discussed and decided upon. That is why the resulting consensus, too, can only be cast temporarily, and why it consists only of the shared understanding of those in the room. An intersubjective understanding, in the terminology of a sociologist, is necessary to jointly decide on the allocation of the available funding; it is stable only for the duration of the meeting, and it is always related to the relative mix of people present and those proposals subject to the discussion. It disintegrates the second the group dissolves and the decision is made, and it has to be summoned anew the next time by working towards posing the same fundamental question again.

I thought long about the notion of a 'black box', which is usually applied to characterize a panel meeting, and I would object that the metaphor is inappropriate. Whenever I sat in one, a panel meeting was more of a specifically designed, highly artificial complex in which particles are manipulated in such a way that they create something like an energetic surplus. Yet this energetic surplus evaporates the minute the specific conditions are dissolved (and the doors of the panel meeting are opened). Maybe it is because of this fragile temporality that the moment is deeply satisfying and almost magical. To be part of it has an elevating, almost revealing character. Everyone in the room is somehow caught by the idea that he or she is more than the sum of the expert knowledge assembled in this room. Everyone is upbeat. I have witnessed it over and over again, and even to me, as a quiet observer at the back, not even having read the proposals, it gave shivers every single time – the thrill of anticipating it, the intellectual satisfaction of having reached consensus, the utilitarian hope of now being able to process smoothly and without any hiccups. While the funding machinery is steaming on, scientific reasoning has achieved another precarious and temporary triumph.

APPENDIX 1
ARCHIVAL COLLECTIONS

Name	Location	Tag	Collection
Anastasia Andrikopoulou	Brussels	EC-AA	Personal collection of files from the European Commission
Archive of European Integration	Pittsburgh	AEI	Online collection of EU-related files
Dan Brändström	Stockholm	DB	Personal files
Community Research and Development Information Service	Brussels	Cordis	Online repository of EU research policy statements
Luc van Dyck	Munich	LvD	Personal collection of files from the ELSF and ISE
European Union Press Release Database	Brussels	Rapid	Online database of press releases and speech transcripts
Pavel Exner	Prague	PE	Personal collection of files from the ERC Scientific Council
Mogens Flensted-Jensen	Copenhagen	MFJ	Personal files
Thomas König	Vienna	TK	Personal collection of files from the ERC Scientific Council
Nederlands Interuniversitair Kunsthistorisch Instituut	Florence	NIKI	Material on trilateral Meeting in Nov. 2004
Alejandro Martin-Hobdey	Brussels	AMH	Personal collection of files from the European Commission
Ramon Marimon Sunyol	Florence	RMS	Personal collection of files from the ERC and the European Commission (via website)
Ernst-Ludwig Winnacker	Strasbourg	ELW	Personal files

APPENDIX 2 INTERVIEWS

Name	Place	Date
Mogens Flensted-Jensen	Copenhagen	03.10.13
Vibeke Hein Olsen	Copenhagen	03.10.13
Martin Bohle	Brussels	30.10.13
William Cannell	London	04.11.13
Ian Halliday	London	04.11.13
Alejandro Martin-Hobdey	Brussels	07.11.13
Pavel Exner	Brussels	02.12.13
Theo Papazoglou	Brussels	04.12.13
Luc van Dyck	Munich	04.02.14
Jack Metthey	Brussels	04.02.14
José Manuel Silva Rodríguez	Brussels	04.02.14
Ernst-Ludwig Winnacker	Strasbourg (tel.)	11.02.14
Enric Banda	Copenhagen	23.06.14
José Mariano Gago	Copenhagen	24.06.14
Dan Brändström	Stockholm (tel.)	23.07.14
Bertil Andersson	Singapore (tel.)	06.08.14
Robert-Jan Smits	Brussels (tel.)	04.02.15

NOTES

INTRODUCTION

1 Hortense Powdermaker, *Stranger and Friend: The Way of an Anthropologist* (New York: Norton, 1966), 290.
2 The role of a 'scientific adviser' was invented under the first ERC President, Fotis C. Kafatos; it was first taken by Charalambos (Babis) Savakis, Professor at Crete University, and head of the Biomedical Science Research Center 'Alexander Fleming' in Greece. Savakis was succeeded in late 2008 by Manolis Antonoyiannakis who took a part-time leave from his position as editor of *Physical Review Letters* in New York until early 2010. With Nowotny succeeding Kafatos as President, the position was obtained by me. It was not continued afterwards, as the new President, Jean-Pierre Bourguignon, was now employed full-time in Brussels.
3 For the span of 2010–11, the ERC's work programme required four regular panel sets in place, while for 2013–14, that number had risen to six (not taking into account additional funding calls that employed a different panel composition, but required another set of panellists) – with each set consisting of 25 panels, and each panel consisting of 12–16 panelists, that meant that, instead of roughly 1,300 scientists, there were now more than 2,700 contracted to act as panel members.
4 Thomas König, 'Funding Frontier Research – Mission Accomplished?', *Journal of Contemporary European Research* 11, no. 1 (2015): 124–35.
5 This is why, at least at three crucial junctures, I cannot provide a definite answer of the reasons why the ERC history developed like it did: in Chapter 3, the decision of the European Commission to join the ERC campaign after a few years of providing only lukewarm support for it; in Chapter 5, the change of the Commission leadership in the Directorate General for Research, which may (or may not) have had an impact on the relationship with the ERC Scientific Council; and, in the same chapter, the Commission's approach to the idea to merge the secretary general position with the ERCEA director.
6 The term has been used self-critically in the introduction of J. Merton England, *A Patron for Pure Science: The National Science Foundation's Formative Years, 1945–57* (Washington, D.C.: NSF, 1982), 7.

7 Most notably, these are Ernst-Ludwig Winnacker, *Europas Forschung im Aufbruch* (Berlin: Berlin University Press, 2012); Julio E. Celis and José Mariano Gago, 'Shaping Science Policy in Europe', *Molecular Oncology* 8, no. 3 (2014): 447–57. An important difference to the authors of those accounts is, of course, that I was only a marginal and late-coming part of the evolving ERC story, and I do not claim to have had any impact on how it developed. This gave me, on the other hand, freedom to look at the ERC with more distance.

8 Hayden White, 'The Question of Narrative in Contemporary Historical Theory', *History and Theory* 23, no. 1 (1984): 33.

9 Helen Wallace and William Wallace, 'Overview: The European Union, Politics and Policy-Making', in *Handbook of European Union Politics*, ed. Knud Erik Jørgensen, Mark Pollack, and Ben Rosamond (London: Sage, 2007), 347.

10 The distinction between science for policy and policy for science seems to stem from Pierre Piganiol, 'Scientific Policy and the European Community', *Minerva* 6, no. 3 (1968): 355; cf. also John Pulparampil, *Science and Society: A Perspective on the Frontiers of Science Policy* (Delhi: Concept, 1978), 43–5.

1 THE FUTURE OF SCIENTIFIC RESEARCH IN EUROPE

1 Janez Potočnik, '[to Scientific Council Members]', Letter (30 June 2005) [EC-AA].

2 Janez Potočnik, 'Speaking Points to the Members of the ERC Scientific Council, on the Occasion of Their First Meeting on 18 October 2005', Unpublished Document (Brussels, 18 October 2005) [EC-AA].

3 Paula E. Stephan, *How Economics Shapes Science* (Cambridge, MA.: Harvard University Press, 2012), 8.

4 If no other source is mentioned, this and the following three paragraphs refer to the recent investigations of Benoît Godin, cf. specifically 'In the Shadow of Schumpeter: W. Rupert Maclaurin and the Study of Technological Innovation', *Minerva* 46, no. 3 (2008): 343–60; *The Making of Science, Technology and Innovation Policy: Conceptual Frameworks as Narratives, 1945–2005* (Montréal: Centre Urbanisation Culture Société de l'INRS, 2009); 'The Linear Model of Innovation: Maurice Holland and the Research Cycle', *Social Science Information* 50, no. 3–4 (2011): 569–81; '"Innovation Studies": The Invention of a Specialty', *Minerva* 50, no. 4 (2012): 397–421; '"Innovation Studies": Staking the Claim for a New Disciplinary "Tribe"', *Minerva* 52, no. 4 (2014): 489–95.

5 Edward Ames, 'Research, Invention, Development and Innovation', *The American Economic Review* 51, no. 3 (1961): 370–81; Fritz Machlup, *The Production and Distribution of Knowledge in the United States* (Princeton, N.J.: Princeton University Press, 1962).

6 Godin, 'Innovation Studies', 493.

7 Vannevar Bush, *Science, the Endless Frontier: A Report to the President* (US Govt. print. off., 1945), 78–9.

8 Only recently, historical studies have begun examining the historical traits and emergence of some of those powerful labels; cf. Désirée Schauz, 'What Is

Basic Research? Insights from Historical Semantics', *Minerva* 52, no. 3 (2014): 273–328; Benoît Godin, 'Innovation: A Study in the Rehabilitation of a Concept', *Contributions to the History of Concepts* 10, no. 1 (2015): 45–68.

9 For a US-centric analysis, cf. David H. Guston, *Between Politics and Science: Assuring the Integrity and Productivity of Research* (Cambridge, MA: Cambridge University Press, 2000); Daniel R. Sarewitz, 'Does Science Policy Matter?', *Issues in Science and Technology* 23, no. 4 (2007).

10 For a nuanced account on the changing patterns in research policy, cf. Dietmar Braun, 'Lasting Tensions in Research Policy-Making — a Delegation Problem', *Science and Public Policy* 30, no. 5 (2003): 309–21; a good overview of recent theories on knowledge production is given by Laurens K. Hessels and Harro van Lente, 'Re-Thinking New Knowledge Production: A Literature Review and a Research Agenda', *Research Policy* 37, no. 4 (2008): 740–60.

11 For example, see Veera Mitzner, 'Research for Growth? The Contested Origins of European Union Research Policy (1963–1974)' (Dissertation, European University Institute, 2013); *Green Paper on Innovation* (Commission of the European Communities: COM/1995/688, 20 December 1995).

12 Susana Borrás and Claudio M. Radaelli, 'The Politics of Governance Architectures: Creation, Change and Effects of the EU Lisbon Strategy', *Journal of European Public Policy* 18, no. 4 (2011): 466.

13 Potočnik, 'Speaking Points to Scientific Council.'

14 The notion is used also in an official Commission document, see *Europe and Basic Research* (European Commission: COM/2004/9, 14 January 2004), 13.

15 'Annual Reports Concerning the Financial Year 2005' (Luxembourg: European Court of Auditors, 31 October 2006), 134.

16 *Towards a European Research Area* (European Commission: COM/2000/6, 18 January 2000), 7, 16, 19.

17 Hans Wigzell, 'Framework Programmes Evolve', *Science* 295, no. 5554 (2002): 445.

18 The quote is of Commissioner Busquin, predecessor of Potočnik, as noted in Olle Edqvist, 'Notes from the Meeting with National Representatives', (Brussels: ERCEG, 14 October 2003), 8 [MFJ].

19 *Decision of 19 December 2006 Concerning the Specific Programme: Ideas Implementing the Seventh Framework Programme of the European Community for Research, Technological Development and Demonstration Activities (2007 to 2013)* (Council of the European Union: EC/2006/972 [O.J. L 400/243], 19 December 2006), 254; the same points had already been brought up in Potočnik, 'Letter.'

20 Cf. Commissioner Máire Geoghegan-Quinn, quoted in 'Annual Report on the ERC Activities and Achievements in 2012' (Brussels: ERC ScC, 2013), 16; 'ERC President Underlines Need for Frontier Research at European Parliament Hearing', Press Release (Brussels: ERC ScC, 13 February 2014); 'On the European Research Council's Operations and Realization of the Objectives Set out in the Specific Programme "Ideas" in 2013' (Brussels: European Commission, 25 August 2014): 8.

21 'In His Own Words: Barroso's Innovation Scorecard', *ScienceBusiness*, 9 October 2014.
22 The notion of 'European added-value' will be examined in greater detail in Chapter 3.
23 Cf. Maria Nedeva, 'Between the Global and the National: Organising European Science', *Research Policy* 42, no. 1 (2013): 220–30; Åse Gornitzka and Julia Metz, 'Dynamics of Institution Building in the Europe of Knowledge: The Birth of the European Research Council', in *Building the Knowledge Economy in Europe: New Constellations in European Research and Higher Education Governance*, ed. Meng-Hsuan Chou and Åse Gornitzka (Cheltenham: Edward Elgar Publishing, 2014), 81–110; Tim Flink, 'Begriffspolitik Europäischen Regierens: Frontier Research und die Entstehung des Europäischen Forschungsrates' (Dissertation, Bielefeld, 2015); Dietmar Braun, 'Actor Constellation in the European Funding Area', in *Towards European Science: Dynamics and Policy of an Evolving European Research Space*, ed. Linda Wedlin and Maria Nedeva (Cheltenham: Edward Elgar Publishing, 2015), 61–82.'
24 Ben Rosamond, 'European Integration and the Social Science of EU Studies: The Disciplinary Politics of a Subfield', *International Affairs* 83, no. 1 (2007): 236; Niilo Kauppi, 'The Political Ontology of European Integration', *Comparative European Politics* 8, no. 1 (2010): 19–36; Ann Zimmermann and Adrian Favell, 'Governmentality, Political Field or Public Sphere? Theoretical Alternatives in the Political Sociology of the EU', *European Journal of Social Theory* 14, no. 4 (2011): 489–515.
25 Thomas F. Gieryn, *Cultural Boundaries of Science: Credibility on the Line* (Chicago: University of Chicago Press, 1999); Pierre Bourdieu, *Homo academicus* (Frankfurt a.M.: Suhrkamp, 1992); Andrew Barry, 'Technological Zones', *European Journal of Social Theory* 9, no. 2 (2006): 239–53; Karin Knorr-Cetina, *Epistemic Cultures: How the Sciences Make Knowledge* (Cambridge, MA: Harvard University Press, 1999).
26 'FY 2014 Agency Financial Report' (Washington, D.C.: NSF, 2014); *Proposal and Award Policies and Procedures Guide* (Washington, D.C.: NSF, 26 December 2014).
27 '2013 Annual Activity Report' (Brussels: ERCEA, 1 July 2014).
28 This does not include various support actions, and calls for the 'Proof-of-Concept' top-up scheme introduced in 2012; those calls amounted to less than one per cent of the ERC's annual operative budget.
29 The agency also spent about € 10.1 million for reviewers, cf. *2013* 'Annual Activity Report', 22.
30 John Ziman, *Real Science: What It Is and What It Means* (Cambridge: Cambridge University Press, 2000), 246.
31 This has excited academics and led to numerous studies on 'intermediary' or 'boundary organizations' (to use political scientists' more preferred lingo for those beasts). Cf. Dietmar Braun and David H. Guston, 'Principal-Agent Theory and Research Policy: An Introduction', *Science and Public Policy* 30, no. 5 (2003): 302–8.
32 Rolf G. Beutel et al., *Insect Morphology and Phylogeny: A Textbook for Students of Entomology* (Berlin: de Gruyter, 2013), 1–103.
33 According to Godin, the term originated in the OECD in the 1990s, and since then developed into a powerful framework for science policy; cf. Godin,

Making; for an empirical comparison of several of those 'systems', see Benedetto Lepori et al., 'Comparing the Evolution of National Research Policies: What Patterns of Change?', *Science and Public Policy* 34, no. 6 (2007): 372–88.

34 Perry Anderson, *The New Old World* (London; New York: Verso, 2009), 132–3.

2 A RADICAL PROPOSAL

1 For a brief overview, see Gunnar Tan, 'Why Do We Have a European Research Council?' (MA Thesis, University of Salzburg, 2010), 60–1.

2 'Lisbon Strategy', Presidency Conclusions (Lisbon: European Council, 23 March 2000).

3 Luca Guzzetti, 'The "European Research Area" Idea in the History of Community Policy-Making', in *European Science and Technology Policy: Towards Integration or Fragmentation*?, ed. Henri Delanghe, Ugur Muldur, and Luc Soete (Cheltenham: Edward Elgar Publishing, 2009), 73–4.

4 Maria João Rodrigues, *European Policies for a Knowledge Economy* (Cheltenham: Edward Elgar Publishing, 2003).

5 Cf. on each of those terms, in consecutive order, Ulrich Beck, *What Is Globalization*? (Cambridge: Polity, 2000); Bruno Amable, Rémi Barré, and Robert Boyer, *Les systèmes d'innovation à l'ère de la globalisation* (Paris: Economica, 1997); Michael Gibbons et al., *The New Production of Knowledge: The Dynamics of Science and Research in Contemporary Societies* (London: Sage, 1994).

6 Trine Flockhart, 'Europeanization or EU-Ization? The Transfer of European Norms across Time and Space', *Journal of Common Market Studies* 48, no. 4 (2010): 787–810.

7 Antonio Ruberti and Michel André, *Un Espace européen de la science: réflexions sur la politique européenne de recherche* (Paris: Presses Universitaires de France, 1995); also, among others, Luca Guzzetti, 'The Development of the Bases for a Community Science and Technology Policy in the Early Seventies', in *History of European Scientific and Technological Cooperation*, ed. John Krige and Luca Guzzetti (Luxembourg: Office for Official Publications of the European Communities, 1997), 411–19; Antonio Ruberti and Michel André, 'The European Union: Science and Technology', *Technology in Society* 19, no. 3–4 (1997): 325–41; Luca Guzzetti, *A Brief History of European Union Research Policy* (Luxembourg: Office for Official Publications of the European Communities, 1995); John Krige and Luca Guzzetti, eds., *History of European Scientific and Technological Cooperation* (Luxembourg: Office for Official Publications of the European Communities, 1997); John Peterson and Margaret Sharp, *Technology Policy in the European Union* (New York: St. Martin's Press, 1998); Susana Borrás, 'Science, Technology and Innovation in European Politics' (Roskilde: Roskilde University, 2000); Stefan Kuhlmann, 'Future Governance of Innovation Policy in Europe – Three Scenarios', *Research Policy* 30, no. 6 (2001): 953–76.

8 Those foresight studies go back to the late 1970s; cf. Rafael Schögler,

'European Union Research Funding. Priority Setting in the Social Sciences and Humanities' (Dissertation, University of Graz, 2013), 39; for projects funded in the 1990s, see 'Targeted Socio-Economic Research Programme. Synopsis of TSER Projects as a Result of the Three Calls for Proposals (1995/96–1997/98)' (Brussels: European Commission, 1999); 'European Perspectives on Science and Technology Policy–Preliminary Outcomes from Policy Research and Debates Generated by the STRATA Projects' (1999–2002)' (Brussels: European Commission, 2003).

9 'STRATA Projects', 23.

10 Paraskevas Caracostas and Luc Soete, 'The Building of Cross-Border Institutions in Europe: Towards a European System of Innovation?', in *Systems of Innovation: Technologies, Institutions, and Organizations*, ed. Charles Edquist (London; Washington: Pinter, 1997), 397.

11 'Second European Report on S&T Indicators' (Brussels: European Commission, 1997); similarly, Yves Farge, 'Industrial Research in the European Union', *Technology in Society* 19, no. 3–4 (1997): 354–55.

12 *ERA*, 4.

13 Guzzetti, *Brief History*; Krige and Guzzetti, *History of European Cooperation*; Mitzner, 'Research for Growth.'

14 Guzzetti, 'Development', 412.

15 Michel André, 'L'espace européen de la recherché: histoire d'une idée', *Journal of European Integration History* 12, no. 2 (2006): 134.

16 'Community Science Policy: New Framework Programme', (Brussels: Commission of the European Communities, June 1983): 1 [AEI].

17 For a good historical overview, see in particular Borrás, 'Science.'

18 Guzzetti, *Brief History*, 83.

19 As it was stated in an early text, the FP was also 'to stimulate discussion within each ministry on the definition of programmes and priorities for each Member State and in international negotiations with a view to harmonizing the various policies so as to avoid the frequent duplication of work which seriously handicaps European research in comparison to the Americans and Japanese.' 'New Framework Programme [1983]', 3.

20 Schögler, 'European Union Research Funding', 41–2.

21 Already with the first FP, the Commission suggested topics where research should be concentrated, and the member states then selected, cf., *Framework Programme for Community Scientific and Technical Activities 1984–1987: First Outline* (Commission of the European Communities: SEC/1982/896, 2 June 1982): 2–5.

22 *Guidelines for a New Community Framework Programme of Technological Research and Development 1987–1991* (Commission of the European Communities: COM/1986/129, 17 March 1986) [AEI].

23 Guzzetti, *Brief History*, 82.

24 *The Maastricht Treaty. Provisions Amending the Treaty Establishing the European Economic Community with a View to Establishing the European Community* (Council of the European Union: [O.J. C 191/01], 1992): Art. 130 i.

25 Pierre Lascoumes and Patrick Le Gales, 'Introduction: Understanding Public Policy through Its Instruments—From the Nature of Instruments to the Sociology of Public Policy Instrumentation', *Governance* 20, no. 1 (2007): 1–21.

26 'Second European Report on S&T Indicators', 8; similarly, Paraskevas Caracostas and Ugur Muldur, *Society, the Endless Frontier: A European Vision of Research and Innovation Policies for the 21st Century* (Brussels: Commission of the European Communities, 1998), 112–25.
27 Thomas Banchoff, 'Institutions, Inertia and European Union Research Policy', *Journal of Common Market Studies* 40, no. 1 (2002): 9.
28 *Stimulating the Community's Scientific and Technological Potential. Experimental Phase: 1983* (Commission of the European Communities: COM/1982/493, 4 August 1982): 2 [AEI].
29 Dan Andrée, 'Priority-Setting in the European Research Framework Programmes' (Stockholm: Vinnova, 2009), 16–20.
30 Paraskevas Caracostas and Ugur Muldur, 'The Emergence of a New European Union Research and Innovation Policy', in *Research and Innovation Policies in the New Global Economy: An International Comparative Analysis*, ed. Philippe Laredo and Philippe Mustar (Cheltenham: Edward Elgar Publishing, 2001), 181–2; Caracostas and Muldur, *Society, the Endless Frontier*, 38.
31 Evaluations in general grappled with the difficulty of assessing the impact of the FP format on the competitiveness of European companies, cf. Terttu Luukkonen, 'The Difficulties in Assessing the Impact of EU Framework Programmes', *Research Policy* 27, no. 6 (1998): 599–610.
32 *Research after Maastricht: An Assessment, a Strategy* (Commission of the European Communities: SEC/1992/682, 9 April 1992): 19, 22 [AEI].
33 Etienne Davignon, 'EU Research and Technological Development Activities. 5-Year Assessment of the European Community RTD Framework Programmes' (Brussels: European Commission, 1997), 7 [AEI].
34 Joan Majo, 'Five-Year Assessment of the European Union Research and Technological Development Programmes, 1995–1999' (Brussels: European Commission, 2000), ii. This report also emphasized the need to go beyond the FP format and to coordinate better R&D policy at European level.
35 For example, in preparation of the fifth edition of the FP, it was proclaimed that science policy now had to enter its 'third phase', 'the marriage of society and innovation' (following phase one after WW2, focusing on basic sciences, and phase two of the 1970s and 80s, concerned about key technologies); cf. Caracostas and Muldur, *Society, the Endless Frontier*, 17, 21.
36 'In form as in content the Sixth Framework Programme will have to be thoroughly rethought out in the light of the project to develop the European research area.' *ERA*, 23.
37 John Finney, 'EuroScience Press Coverage', *Euroscience News*, September 1997.
38 Alison Abbott, 'European biologists unite to lobby for more money', *Nature*, 401, no. 6756 (28 October 1999): 834; ELSF was a roof organization of life science associations, such as the European Life Scientist Organization (ELSO), European Molecular Biology Laboratory (EMBL), European Molecular Biology Organization (EMBO), and the Federation of European Biochemical Societies (FEBS), with the aim 'to provide a coherent voice for the many different European societies representing subsections of the Biosciences. This coalition ensures that the positions of life scientists are clearly enunciated.' Cf. Frank Gannon and Luc van Dyck, 'European

Research Council – the Life Scientist's View. A Document from the European Life Sciences Forum Based on Consultations with the Life Scientists' Community' (Heidelberg: ELSF, October 2003), 19 [LvD].

39 Cf. Lennart Philipson, 'Integration of European Life Sciences', *Science* 256, no. 5056 (24 April 1992): 478; Wilhelm Krull and Frieder Meyer-Krahmer, 'Science, Technology, and Innovation in Germany – Changes and Challenges in the 1990s', in *Science and Technology in Germany*, ed. Wilhelm Krull and Frieder Meyer-Krahmer (London: Cartermill, 1996), 24; Jean-Patrick Connerade, 'European Science Policy: Just Where Is It Going?', *Euroscience News*, January 2000; also Tan, 'European Research Council', 62.

40 Virginia Acha, Orietta Marsili, and Richard Nelson, 'What Do We Know about Innovation?', *Research Policy* 33, no. 9 (2004): 1253–58.

41 Keith Pavitt, 'The Inevitable Limits of EU R&D Funding', *Research Policy* 27, no. 6 (1998): 559–68.

42 'Lisbon Strategy.'

43 Keith Pavitt, 'Why European Union Funding of Academic Research Should Be Increased: A Radical Proposal', *Science and Public Policy* 27, no. 6 (2000): 459.

44 Ibid., 455, 457, 458.

45 Ibid., 459.

46 'America's Basic Research: Prosperity through Discovery. A Policy Statement' (New York; Washington, D.C.: Committee for Economic Development, 1998); Nathan Rosenberg, *Schumpeter and the Endogeneity of Technology: Some American Perspectives* (London; New York: Routledge, 2000). Pavitt quoted from this book in the beginning of his article. It was based on a lecture series held by Rosenberg in 1998 at the University of Graz.

47 Pierre Papon, 'The Role of Research Infrastructures in the Emergence of European Science: Present and Future', in *European S&T Policy and the EU Enlargement. Workshop of experts from pre-accession CEEC and EUROPOLIS Project Group*, ed. Simeon Anguelov and Pierre Lasserre (Venice: UNESCO, 2000), 82.

48 Peter Tindemans and Pierre Papon, 'New Funding Mechanisms for European Basic and Strategic Science' (Strasbourg: EuroScience, 10 April 2001).

49 The Swedish government, at this point, was not putting basic research on its list of priority during the presidency, cf. the interview with Swedish research minister Thomas Östros, 'Östros outlines Research Council agenda', Interview (Cordis: 16328, 27 February 2001).

50 Pär Omling and Uno Svedin, 'Summary of the Krusenberg Manor Workshop' (Stockholm: Swedish Research Council, 25 April 2001) [MFJ].

51 Mogens Flensted-Jensen, 'Mathematics and Research Policy – A View Back on Activities I Was Happy to Take Part in' (Copenhagen, 3 May 2013), 4 [MFJ].

52 '2000 Annual Report on the Socio-Economic Dimension in the Fifth Framework Programme' (Brussels: European Commission, 2001), 51 [AEI].

53 Dan Brändström, 'Styrelsen för Stiftelsen Riksbankens Jubileumsfonds berättelse över fondens verksamhet och förvaltning under år 2001' (Stockholm: Riksbankens Jubileumsfond, 2001), 31 [DB] [translated by TK].

54 Ibid., 32.
55 'The Bank of Sweden Tercentenary Foundation Annual Report 2002' (Stockholm: Riksbankens Jubileumsfond, 2002), 40–1.
56 Stefan Einarsson, 'Sweden: Exploratory Overview of Research Foundations', in *Understanding European Research Foundations. Findings from the FOREMAP Project* (London: Alliance Publishing Trust, 2009), 95–120.
57 For an overview, cf. Mats Benner and Ulf Sandström, 'Inertia and Change in Scandinavian Public-Sector Research Systems: The Case of Biotechnology', *Science and Public Policy* 27, no. 6 (2000): 449; Tomas Hellström and Merle Jacob, 'Taming Unruly Science and Saving National Competitiveness: Discourses on Science by Sweden's Strategic Research Bodies', *Science, Technology & Human Values* 30, no. 4 (2005): 443–67.
58 Pär Omling, 'Sweden's Research Shakedown Provides Clearer Vision', Interview (Cordis: 16859, 30 May 2001).
59 Uno Lindberg, 'Experts Demand a New Science Policy for Europe', Press Release (Stockholm: Royal Swedish Academy of Sciences, 22 April 2002); similarly Peter Tindemans and Pierre Papon, 'The European Research Area: A New Frontier and Scenario for Research in Europe. A Comparative Analysis Based on the Scenarios and the Recommendations of the Europolis Project' (Brussels: European Commission, June 2002).
60 Jørgen Søndergaard and Mogens Flensted-Jensen, 'Towards a European Research Area. Do We Need a European Research Council?' (Copenhagen: Danish Research Councils, November 2002), 4.
61 'European Council for Research Still a Decade Away, Says Leading Academic', Press Release (Cordis: 18128, 15 March 2002).
62 George Radda, 'Biomedical Research and International Collaboration', *Science* 295, no. 5554 (2002): 445–6.
63 'High-Level Panel Split over the Need for a European Research Council', Press Release (Cordis: 19235, 13 November 2002).
64 Luc van Dyck, 'Footing the Bill', *EMBO Reports* 4, no. 6 (1 June 2003): 544.
65 Particularly the European Research Advisory Board under Helga Nowotny's direction followed this approach, cf. 'EURAB Report of Activities 2001–2003' (Brussels: EURAB, 2003), 15–6.
66 Quirin Schiermeier, 'European Research Council: A Window of Opportunity', *Nature* 419, no. 6903 (12 September 2002): 108–9.
67 Ernst-Ludwig Winnacker, 'European Science', *Science* 295, no. 5554 (2002): 446.
68 Enric Banda, 'Science in Europe. Implementing the European Research Area', *Science* 295, no. 5554 (2002): 443.
69 'A European Research Council—More Competition in Science [Interview with Enric Banda]', *EMBO Reports* 3, no. 4 (1 April 2002): 292–5.
70 'New Research Council Needed', *Nature* 418, no. 6895 (18 July 2002): 259.
71 'ESF, the next Decade: A Reappraisal of ESF's Strategic Mission' (Strasbourg: European Science Foundation, 1993).
72 'ESF Report Gives New Impetus to Debate on Establishment of a European Research Council', Press Release (Cordis: 20190, 5 May 2003).
73 Craig Venter and Daniel Cohen, 'The Century of Biology', *New Perspectives Quarterly* 14, no. 5 (1997): 26–31.

74 Michael Ashburner, 'Europe Must Grant Crucial Funds for Biological Research', *Nature* 402, no. 6757 (4 November 1999): 12–22; Fotis C. Kafatos, 'Interesting Times-Biology, European Science, and EMBL', *Science* 287, no. 5457 (25 February 2000): 1401–3.
75 This was explicitly mentioned during the first of a series of conferences organized by ELSF in the aftermath of the Copenhagen conference, cf. Frank Gannon, 'Summary of the Meeting 'Life Sciences in the European Research Council – The Scientists' Opinion,' (Paris: ELSF, 19 February 2003), 3 [LvD].
76 Those concerns were nicely summarized by Winnacker's successor as head of Eurohorcs, cf. Rolf Tarrach, 'Contribution to Paris Meeting', Talking Points (Paris, 19 February 2003) [LvD].
77 Winnacker, 'European Science'; similarly his position in Schiermeier, 'European Research Council', 109.
78 Richard Sykes, 'New Structures for the Support of High-Quality Research in Europe. A Report from a High Level Working Group Constituted by the European Science Foundation to Review the Option of Creating a European Research Council', (Strasbourg: European Science Foundation, April 2003), 15.
79 Ibid.
80 Søndergaard and Flensted-Jensen, 'Towards', 13.
81 This was pointed out most clearly in an article by Wilhelm Krull (who would later be part of the ESF working group), cf. Wilhelm Krull, 'A Fresh Start for European Science', *Nature* 419, no. 6904 (19 September 2002): 250.
82 'More Competition in Science [Interview with Enric Banda]', 294; David J. v. H Grønbæk, 'A European Research Council: An Idea Whose Time Has Come?', *Science and Public Policy* 30, no. 6 (2003): 399.
83 Cf. Sykes, 'New Structures', 13.
84 Cf. Grønbæk, 'Idea', 402.
85 Søndergaard and Flensted-Jensen, 'Towards', 21.
86 Ibid., 13.
87 Luc van Dyck, 'Follow-up Meeting Life Sciences and the European Research Council. Concrete Proposals Concerning Grants, Infrastructure and Delivery Mechanisms' (Venice: ELSF, 29 May 2003), 11 [LvD].
88 Carol Featherstone and Kai Simons, 'Outline for a European Research Council', *Nature* 425, no. 6957 (2 October 2003).
89 Søndergaard and Flensted-Jensen, 'Towards', 5.
90 'European Research Council' (Amsterdam: ALLEA, 7 November 2002) [EC-AA]; Gannon, 'Summary'; Sykes, 'New Structures'; 'Eurohorcs Declaration on Reinforced Research Cooperation in Europe' (Oegstgeest: Eurohorcs, 19 May 2003) [EC-AA].
91 'European Research Council', Press Release (Brussels: EURAB, 1 November 2002).
92 Grønbæk, 'Idea', 395.
93 Ibid., 396.
94 Ibid., 402.
95 This was also implicitly acknowledged when van Dyck mentioned that '[t]he scientific community responded with interest and enthusiasm to the concept of an ERC, albeit in a rather dispersed manner'; cf. Luc van Dyck,

'European Research Council. An Initiative for Science in Europe' (Brussels: European Parliament, 23 February 2004) [LvD].

3 EUROPEAN VALUE ADDED

1 Richard Stone, 'Europe Begins Work on Modest New Agency', *Science* 296, no. 5569 (2002): 827.
2 '2467th Meeting of the Council of the European Union Competitiveness held in Brussels on 26 November 2002', Draft Minutes (ST 14815 2002 INIT, 5 December 2002): 6; similarly, '2467th Council Meeting – Competitiveness', Press Release (ST 14365 2002 INIT, 26 November 2002): 20.
3 Luc van Dyck, 'Unique New Platform Offers European Scientific Community a Common Voice', Press Release (Cordis: 102475, 28 October 2004).
4 Only the EU presidencies for 2004 had the ERC on their agenda again; the Dutch had even devised a report on its feasibility already, cf. Peter A. Schregardus and Gerard J. Telkamp, 'Towards a European Research Council? Study Commissioned by the Netherlands Ministry of Education, Culture and Science' (The Hague: The Netherlands Institute of International Relations 'Clingendael', 9 September 2002) [EC-AA].
5 Sander's letter is partly recited in Federico Mayor, 'The European Research Council. A Cornerstone in the European Research Area' (Copenhagen: Ministry of Science, Technology and Innovation, December 2003), 33–4 [MFJ]. I am grateful to Vibeke Hein Olsen, then working at the office of the Danish Research Councils and one of Flensted-Jensen's closest collaborators, for a vivid recollection of the events in December 2002 (see Appendix 2).
6 'The Bank of Sweden Tercentenary Foundation Annual Report 2003' (Stockholm: Riksbankens Jubileumsfond, 2003), 44.
7 The ERCEG members all knew each other for a long time. Nowotny, Kroó, Megie were members of EURAB; Nowotny had known Brändström from the time he commissioned her to do a study on the Swedish research landscape (which would then become the path-breaking study by Gibbons et al., *New Production of Knowledge*); Kroó and Krull both sat on the Governing Board of EuroScience; and so on.
8 Mayor, 'Cornerstone', 3.
9 Kai Simons and Carol Featherstone, 'The European Research Council on the Brink', *Cell* 123, no. 5 (2 December 2005): 748; Celis and Gago, 'Shaping Science Policy in Europe', 452.
10 Philippe Busquin, 'Towards a European Research Area. Do We Need a European Research Council?', Unpublished Document (Copenhagen, 7 October 2002), 5 [EC-AA]; similarly, Gannon, 'Summary', 4; 'Politicians, Scientists and Officials Begin to Give Shape to ERC', Press Release (Cordis: 19760, 21 February 2003).
11 *The European Research Area: Providing New Momentum. Strengthening – Reorienting – Opening up New Perspectives* (European Commission: COM/2002/565, 16 October 2002): 20.
12 Banchoff, 'Institutions', 17.
13 The change in tone was not lost to the members of the Mayor group; cf. Flensted-Jensen, 'Mathematics', 5.
14 Mayor, 'Cornerstone', 24.

15 Ibid., 5, 14.
16 Søndergaard and Flensted-Jensen, 'Towards', 5.
17 Mayor, 'Cornerstone', 24.
18 Symbolically, this was done during a meeting with Nobel Laureates; 'Commission Answers Calls for a European Research Council', Press Release (Cordis: 21019, 9 October 2003); around the same time, at the ELSF meeting in Dublin, Mitsos 'announced that the Commission was supporting the idea of an instrument for basic research and, for this purpose, would request a specific credit line from the EU budget'; see van Dyck, 'ERC', 4.
19 Edqvist, 'Notes', 7; similarly, Mitsos stressed 'that the Commission was supporting the idea of an instrument for basic research and, for this purpose, would request a specific credit line from the EU budget', cf. Luc van Dyck, 'A European Research Council for All Sciences. 21–22 October 2003' (Dublin: Irish Royal Academy, January 2004), 4 [LvD].
20 'Irish Minister Calls for New Initiative to Support Basic Research, but without the Red Tape', Press Release (Cordis: 21608, 17 February 2004); 'Dutch Ministers Quietly Confident about Achieving Presidency Priorities', Press Release (Cordis: 22288, 6 July 2004).
21 For a general overview of the procedures, cf. Mark Pollack, 'Delegation and Discretion in the European Union', in *Delegation and Agency in International Organizations*, ed. Darren G. Hawkins et al. (Cambridge: Cambridge University Press, 2006), 165–96; for an account on the seventh edition of the FP format, Ugur Muldur et al., *A New Deal for an Effective European Research Policy: The Design and Impacts of the 7th Framework Programme* (Dordrecht: Springer, 2006), 224–254.
22 Rolf Linkohr, 'EP Rapporteur Calls for Speedy Implementation of Commission's Action Plan on Research Investment', Press Release (Cordis: 20860, 9 September 2003).
23 '2525th Council Meeting – Competitiveness', Press Release (Council of the European Union: ST 12339 2003 INIT, 22 September 2003): 32; in the interview (cf. Appendix 2), Brändström explicitly mentioned that this was triggered by Michael Sohlman and Étienne-Émile Baulieu, then the President of the French Académie des Sciences, who intervened with the French Minister.
24 Mayor, 'Cornerstone', 4–5.
25 Gannon, 'Summary.'
26 'There was a discussion of the size of budget needed for the ERC. Someone suggested, that we should ask for 500 million Euro's a year. But the whole group reacted, that if the ambition was so low, we could just as well stop the work immediately. After a call to the Commission Peter Kind [the observer from the Commission, TK] surprisingly proposed, that we should aim at least at 2 billion Euro's a year.' Flensted-Jensen, 'Mathematics', 5.
27 Achilleas Mitsos, 'A European Research Council – the Commission View', Unpublished Document (Dublin, 22 October 2003) [EC-AA]; the Mayor Report opens with the sentence 'It is now time to bring a new definition of added value.' Mayor, 'Cornerstone', 4.
28 Mitsos, 'Commission View', 4.
29 Armin von Bogdandy, 'An Autonomous ERC – Legal Challenges', Unpublished Document (Brussels: European Parliament, 23 February 2004), 4 [LvD].
30 Kees Van Kersbergen and Bertjan Verbeek, 'The Politics of Subsidiarity in

the European Union', *Journal of Common Market Studies* 32, no. 2 (1994): 227; for a specific account on the political usage of the term in the preparation of the seventh edition of the FP format, cf. Muldur et al., *New Deal*, 183–223.

31 *Basic Research*, Europe and 10.

32 Ibid., 13.

33 Ibid., 12–13.

34 'Outcome of Proceedings of the Competitiveness Council on 11 March 2004', Council Conclusions (ST 7379 2004 INIT), 2004; 'Informal Competitiveness Council, 1-3.July 2004, Maastricht', Presidency Summary (Maastricht: Competitiveness Council, 3 July 2004) [NIKI].

35 Edqvist, 'Notes', 2.

36 David Sainsbury, 'Dublin Symposium Speaking Note', Unpublished Document (Dublin, 16 February 2004), 4 [EC-AA].

37 'Calls to Establish an ERC May Be Premature, Warns the Royal Society', Press Release (Cordis: 21446, 15 January 2004); the Royal Society soon afterwards published a report with more nuanced undertone, cf. Julia Higgins et al., 'The Place of Fundamental Research in the European Research Area: The Royal Society Response to the Mayor Report' (London: UK Royal Society, March 2004).

38 It probably helped that the Chief Scientific Adviser to the Blair Government worked on a bibliometric report that took a pro-European perspective; cf. David A. King, 'The Scientific Impact of Nations', *Nature* 430, no. 6997 (15 July 2004): 311–16.

39 *Science and Technology, the Key to Europe's Future – Guidelines for Future European Union Policy to Support Research* (European Commission: COM/2004/353, 16 June 2004): 6.

40 'Italy's Contribution to FP7 Debate Raises Doubt over the Need for an ERC', Press Release (Cordis: 22720, 4 October 2004); an internal Commission memo mentioned 'Finland, Italy and Spain' to be 'still reticent', cf. 'Basic Research (The European Research Council) [in View of the Special Session of ITRE]', Briefing Document (Brussels: European Commission, 17 September 2004) [EC-AA].

41 'Italian Contribution to the Debate on the Future of European Research Policy', Unpublished Document (Rome: Ministero dell'Istruzione dell'Università e della Ricerca, 2004), 8 [NIKI].

42 'Discussion Paper Florence Seminar', Briefing Document (Brussels: European Commission, 28 October 2004), 3 [EC-AA].

43 Alexander Tenenbaum, 'The European Research Council', Unpublished Document (Rome: Ministero dell'Istruzione dell'Università e della Ricerca, October 2004), 2 [TK].

44 Ibid.

45 '2624th Council Meeting – Competitiveness', Draft Minutes (Council of the European Union: ST 15259 2004 INIT, 25–6 December 2004): 22.

46 Ibid., 19.

47 *Proposal for a Decision of the European Parliament and of the Council Concerning the Seventh Framework Programme of the European Community for Research, Technological Development and Demonstration Activities (2007 to 2013)* (European Commission: COM/2005/119 [O.J. C 125/11], 6 April 2005): 3.

48 Based on documents from http://www.consilium.europa.eu/register/en/content/int/?lang=EN&typ=ADV, accessed 15 September 2015.
49 'Greek Position Regarding the Seventh RTD Framework Programme of the EU' (ST 6062 2005 INIT, 4 February 2005), 7.
50 'Austrian Position Regarding the Seventh RTD Framework Programme of the EU' (ST 15242 2004 INIT, 23 November 2004), 9.
51 Martin Enserink, 'Only the Details Are Devilish for New Funding Agency', *Science* 306, no. 5697 (29 October 2004): 796.
52 Cf. André Sapir et al., 'An Agenda for a Growing Europe. Making the EU Economic System Deliver' (Brussels: European Commission, July 2003), 134.
53 William C. Harris et al., 'Frontier Research: The European Challenge' (Brussels: European Commission, February 2005), 18.
54 *Technical Adjustments to the Commission Proposal for the Multiannual Financial Framework 2007–2013* (European Commission: SEC/2005/494, 2005).
55 Some useful reflections on the political games around the MFF can be found in Peter Becker, 'Lost in Stagnation. The EU's Next Multiannual Financial Framework (2014–2020) and the Power of the Status Quo' (Berlin: Stiftung Wissenschaft und Politik, October 2012).
56 More precisely, the proposal was to commit € 72.010 billion to competitiveness as a whole over seven years, in comparison to the € 132.755 billion proposed by the Commission; cf. *Financial Perspective 2007–2013 – Negotiating Box* (Luxembourg EU Presidency: ST 10090 2005 INIT, 15 June 2005), 5.
57 'European Research Budget in Dire Straits', *EurActiv* (13 June 2005).
58 *Financial Perspective 2007–2013 – Final Comprehensive Proposal* (UK EU Presidency: ST 15915 2005 INIT, 19 December 2005), 5.
59 The fact that Mitsos hinted at the outcome some weeks before the MFF was adopted indicates that there must have been negotiations in the background; cf. 'No Doubling of Research Budget but 'Significant' Increase, Predicts Mitsos', Press Release (Cordis: 24765, 16 November 2005); cf. also 'UK Budget Proposals Provide for 75 per Cent Increase in EU Research Funds', Press Release (Cordis: 24932, 15 December 2005).
60 Fotis C. Kafatos, 'European Research Council. An Initiative for Science in Europe', Presentation (Brussels: European Parliament, 23 February 2004), 4 [LvD].
61 Mayor, 'Cornerstone', 14–15.
62 Setting up the ERC in a member state would have meant to put it under the legal framework of that state and it was doubtful if all other member states would have selflessly contributed to this organization; furthermore, ESF (which had been established under French law) was not a role model to advocates. Establishing an intergovernmental solution, on the other hand, had been done successfully before with research-performing facilities like EMBL and CERN; but it would have required lengthy negotiations and would have taken several years.
63 *Treaty of Amsterdam Amending the Treaty on European Union, the Treaties Establishing the European Communities and Certain Related Acts* (Council of the European Union: [O.J. C 340/01], 2 October 1997), Art. 171.
64 *Council Regulation Laying down the Statute for Executive Agencies to Be Entrusted with Certain Tasks in the Management of Community Programmes*

(Council of the European Union: EC/58/2003 [O.J. L 57/1], 19 December 2002), 2.

65 'Delegating Implementing Tasks to Executive Agencies: A Successful Option?' (Luxembourg: European Court of Auditors, 2009), 13.

66 von Bogdandy, 'Autonomous', 7.

67 Already the Mayor Report had cautioned that this model 'will impose organizational, financial and auditory mechanisms and regulations on the ERC, which seems difficult to combine with the required autonomy.' Mayor, 'Cornerstone', 14–15.

68 This was made explicit in the 'Partnership model' following von Bogdandy's presentation, cf. Armin von Bogdandy and Dietrich Westphal, 'Untersuchung zur Implementierung eines Europäischen Forschungsrates (ERC) im 7. EU-Rahmenprogramm' (Munich: Max Planck Gesellschaft, 11 July 2005), 13–7.

69 von Bogdandy, 'Autonomous', 7.

70 '[Briefing in View of the Special Session of ITRE]', 1.

71 The proposal was presented in September 2005 – as required by the EU treaty, the ERC was packed up in a 'Specific Programme' called 'Ideas'; cf. 'European Commission Develops Its Plans for Future Research Programme', Press Release (Rapid: IP/05/1171, 21 September 2005).

72 Cf. Alan Johnson, 'The Informal Competitiveness Council, Cardiff 11–12 July 2005', in *House of Lords European Union Committee: Correspondence with Ministers, March 2005 to January 2006* (London: Parliament Publications and Records, 21 July 2005), 104.

73 Achilleas Mitsos, 'Closing Address', Unpublished Document (Dublin, 17 February 2004), 5 [EC-AA].

74 Ernst-Ludwig Winnacker, 'A Rocky Start – the Early Days of the European Research Council (ERC)', Unpublished Document (Berlin, February 2013), 21–3 [ELW].

75 'A European Fundamental Research Funding Mechanism: Implementing Principles' (Brussels: European Commission, July 2004) [EC-AA].

76 Ernst-Ludwig Winnacker et al., 'A European Funding Mechanism for Basic Research' (Brussels: European Commission, 28 July 2004), 2 [EC-AA].

77 'Scientific Council of the European Research Council Announced', Press Release (Rapid: IP/05/956, 18 July 2005).

78 According to the summary of the informal Council meeting in July 2005, ministers ' [w]elcomed the work of the European Commission, and that of the Identification Committee chaired by Lord Patten, in identifying the means to foster excellence in basic research, including the possible establishment of a European Research Council in the period of the next Framework Programme and noted that the Commission's proposal would need to effectively guarantee the independence and autonomy of the ERC.' Johnson, 'Informal Competitiveness Council', 106.

79 *Guidelines for Future European Union Policy to Support Research* (European Parliament: P6_TA(2005)0077, 10 March 2005): point 22, based on motion presented by rapporteur Pia Elda Locatelli.

80 Janez Potočnik, 'ERC: Implementing Structures', Briefing Document (Brussels: European Commission, 19 September 2005) [EC-AA].

81 Dan Brändström, 'Result of High-Level Round Table Discussion on the Scientific Autonomy and Self-Governance of the European Research Council', Unpublished Document (Brussels, 19 September 2005), 2 [DB].

82 *Draft Decision of the European Parliament and of the Council concerning the Seventh Framework Programme of the European Community for Research, Technological Development and Demonstrative Activities (2007 to 2013)*, Partial General Approach (Council of the European Union: ST 15062 2005 INIT, 28–29 November 2005), 50.
83 '2694th Council Meeting – Competitiveness', Press Release (Council of the European Union: ST 14155 2005 INIT, 28–9 November 2005): 8; see also '2694th Meeting of the European Union (competitiveness) held in Brussels on 28 and 29 November 2005', Minutes (Council of the European Union: ST 15031 2005 INIT REVI, 22 December 2005).

4 THE MOST PROMISING OPPORTUNITIES

1 *National Science Foundation Act of 1950, US Code*, vol. 42, 10 May 1950.
2 David E. Lewis and Jennifer L. Selin, *Sourcebook of United States Executive Agencies* (Washington, D.C: Administratitive conference of the US, 2012), 48–54; cf. also 'The United States Government Manual' (Washington, D.C: office of the Federal Register 2013).
3 See, for example, Jeffrey Mervis, 'Battle between NSF and House Science Committee Escalates: How Did It Get This Bad?', *ScienceInsider* (2 October 2014).
4 Bart Van Ballaert, 'The Politics behind the Consultation of Expert Groups: An Instrument to Reduce Uncertainty or to Offset Salience?', *Politics and Governance* 3, no. 1 (31 March 2015): 139–50.
5 Potočnik, 'Speaking Points to Scientific Council.'
6 Neil Kinnock, 'Reforming the European Commission: Organisational Challenges and Advances', *Public Policy and Administration* 19, no. 3 (1 July 2004): 7–12; Emmanuelle Schön-Quinlivan, 'Implementing Organizational Change – the Case of the Kinnock Reforms', *Journal of European Public Policy* 15, no. 5 (2008): 726–42.
7 Potočnik, 'Speaking Points to Scientific Council.'
8 That does not mean that ISE stopped supporting the ERC; quite the contrary, see Etta Kavanagh, ed., 'Crucial Choices for the Nascent ERC', *Science* 311, no. 5765 (3 March 2006): 1240. But it is important to note that this was now more of an accompaniment to the further development of the ERC. As soon as the Scientific Council had been gathered, ISE was seeking new fields of action, as can be seen from the content of the conference in late 2005, cf. 'Celebrating the First Concrete Steps towards the Implementation of the ERC', Meeting Report (Paris: ISE, 10.11 2005) [LvD].
9 '1st Meeting of the ERC Scientific Council', Minutes (Brussels: ERC ScC Plenary #01, 18 October 2005), 1 [PE].
10 Nowotny's account aimed at reminding newcomers that '[a]t stake is the continuing sense that the ERC belongs to the scientific community.' Cf. Helga Nowotny, 'Preserve the European Research Council's Legacy', *Nature* 504, no. 7479 (11 December 2013): 189.
11 '1st Meeting of the ERC Scientific Council', 2.
12 This anecdote has been told by several members of the Scientific Council (see Appendix 2).
13 Potočnik, 'Speaking Points to Scientific Council.'

14 Chris Patten et al., 'European Research Council Identification Committee' (Brussels: European Commission, 21 March 2005), 3.
15 Ibid., 2; this consideration was probably influenced by an article of Robert May, then president of the Royal Society, cf. 'Raising Europe's Game', *Nature* 430, no. 7002 (19 August 2004): 831–2.
16 Fotis Kafatos, quoted in Janet Fricker, 'Putting Europe on the Scientific Map', *Molecular Oncology* 3, no. 5 (1 December 2009): 388.
17 'Report of the ERC Scientific Council on the Occasion of Its Formal Establishment' (Brussels: ERC ScC 2 February 2007).
18 Maybe cleverly anticipating the value of this symbol, the notion of the 'founding members' also found its way into the legislative text, cf. *Decision Establishing the European Research Council* (European Commission: EC/2007/139 [O.J. L 57/14], 2 February 2007): 15, 19.
19 Three members officially withdrew their membership in 2008; cf. 'ERC: Activities and Achievements in 2008' (Brussels: ERC ScC, 2009), 25. Records show that they were not attending ERC meetings as early as 2007.
20 It didn't help that Castells was later appointed to the board of the European Institute for Innovation and Technology (EIT), a pet project by Commission President Barroso and that was observed suspiciously by the Scientific Council leadership. 'British Academic Appointed to First Governing Board of the European Institute of Technology', Press Release (European Commission: 1P/08/1220).
21 'Procedure for Election of the Chairperson of the SC' (Brussels: ERC ScC Plenary #01, 18 October 2005).
22 'Report Formal Establishment', 1.
23 Fotis C. Kafatos, 'Towards an Integrated ERC' (Brussels: ERC ScC Plenary #01, 24 January 2006), 1.
24 Council of the European Union, *Executive Agencies*.
25 'ERC Press Statement', Press Release (Brussels: ERC ScC Plenary #01, 19 October 2005).
26 The letter is also mentioned in Winnacker, *Aufbruch*, 47.
27 Fotis C. Kafatos, 'Recruitment Process for ERC Secretary-General' (Vienna: ERC ScC Plenary #03, 26 April 2006); '3rd Meeting of the ERC Scientific Council, 26.–27.4.2006', Minutes (Copenhagen: ERC ScC Plenary #04, 29 May 2006), 3.
28 According to Winnacker, his weak performance was due to a sleepless night in a shoddy hotel room; cf. Winnacker, *Aufbruch*, 45.
29 'Recruitment Committee Report' (London: ERC ScC Plenary #05, 24 July 2006).
30 The summary of Kafatos' interviews expressed his favour for Winnacker: 'For me, he is the clear choice. Based on his experience, his match with the requirements that we advertised and need, and his admirable record in a position that is directly relevant to the tasks that we face, I am confident that he will bring an essential operational talent.' Fotis C. Kafatos, 'In-Depth Interviews Final Report' (London: ERC ScC Plenary #05, 4 August 2006), 10.
31 Fotis C. Kafatos and Helga Nowotny, 'Message from Fotis Kafatos and Helga Nowotny' (London: ERC ScC Plenary #05, 16 August 2006).
32 Ibid.
33 'ERC Secretary General Announced', Press Release (Brussels: ERC ScC, 30 August 2006) [PE].

34 Winnacker, *Aufbruch*, 46–7, writes that he was never informed why the position was split between him and Mas-Colell; he suspects that either the social scientists did not trust him, or that the Commission put pressure on the Scientific Council to supplement a Northern European with someone from the South. The social scientists, however, were too small a subgroup within the council to effectively veto anyone from taking office. And, as far as the sources of the Scientific Council are concerned, the Commission had no influence whatsoever on the election process. It was a procedure exclusively set up and run by the Scientific Council. As a matter of fact, the Commission was not particularly happy with the position of the secretary general in the first place.

35 I am grateful to have discussed this in detail with William Cannell (see Appendix 2).

36 The programme was based on the legislative provision to take care of 'scientific and technological needs', cf. *Decision Concerning the Sixth Framework Programme of the European Community for Research, Technological Development and Demonstration Activities, Contributing to the Creation of the European Research Area and to Innovation (2002 to 2006)* (European Parliament and Council of the European Union: EC/2002/1513 [O.J. L 232/1], 27 June 2002): 5.

37 'Anticipating Scientific and Technological Needs. New and Emerging Science and Technology (NEST)', Work Programme (Brussels: European Commission, 2002), 3.

38 *Implementing the Renewed Partnership for Growth and Jobs. Developing a Knowledge Flagship: The European Institute of Technology* (European Commission: COM/2006/77, 22 February 2006).

39 Åse Gornitzka and Julia Metz, 'European Institution Building under Inhospitable Conditions – the Unlikely Establishment of the European Institute of Innovation and Technology', in *Building the Knowledge Economy in Europe: New Constellations in European Research and Higher Education Governance*, ed. Meng-Hsuan Chou and Åse Gornitzka (Cheltenham: Edward Elgar Publishing, 2014), 117; see also Peter D. Jones, 'The European Institute of Technology and the Europe of Knowledge: A Research Agenda', *Globalisation, Societies and Education* 6, no. 3 (2008): 291–307.

40 Helga Nowotny, 'The European Institute of Technology (EIT) – Scope for Guidance from the ERC Scientific Council' (Vienna: ERC ScC Plenary #03, 26 April 2006).

41 '6th Meeting of the ERC Scientific Council, London, 5.–6.10.2006', Minutes (Ljubljana: ERC ScC Plenary #07, 20 November 2006), 1.

42 The ERC President expressed his hopes that, since 'we need to finalize a number of documents [. . .], the restricted configuration of the ERC Board will considerably facilitate decision-making'; cf. Fotis C. Kafatos, 'ERC Board', Discussion Paper (Ljubljana: ERC ScC Plenary #07, 10 November 2006).

43 'ERC Board Meeting, 28.11.2006', Minutes (Brussels: ERC ScC Plenary #08, 11 January 2007), 1.

44 '6th ERC Scientific Council Meeting', Agenda (London: ERC ScC Plenary #06, 10 May 2006), 1.

45 'Minutes 6th Meeting, 5.–6.10.2006.'

46 Cf. 'Point Summary ERC Scientific Council Extended Chairs' Meeting,

28.3.2006', Minutes (Vienna: ERC ScC Plenary #03, 26 April 2006), 1 [PE]; 'Minutes 3rd Meeting, 26–7.4.2006', 1.

47 Metthey was a high-ranking functionary formerly directing 'research actions for transport' under the sixth edition of the Framework Programme, cf. 'New Deputy Director General and Directors for Research DG Announced', Press Release (Cordis: 18001, 15 February 2002).

48 'Directorate S – Mission Statement' (Brussels: ERC ScC Plenary #07, 20 October 2006), 1.

49 This terminology and categorization follows broadly the hierarchy as shown in 'Has the Commission Ensured Efficient Implementation of the Seventh Framework Programme for Research?', (Luxembourg: European Court of Auditors, 2013), 13; specifically on the ERC (although incomplete), see also 'The ERC Legal Base' (Milan: ERC ScC Plenary #08, 11 January 2007).

50 *Laying down the Rules for the Participation of Undertakings, Research Centres and Universities in Actions under the Seventh Framework Programme and for the Dissemination of Research Results (2007–2013)* (European Parliament and Council of the European Union: EC 2006/1906 [O.J. L 391/1], 18 December 2006).

51 'Agenda', 10 May 2006; '7th ERC Scientific Council Meeting', Agenda (Ljubljana: ERC ScC Plenary #07, 20 November 2006).

52 'Minutes 6th Meeting, 5.–6.10.2006', 2.

53 *ERC Decision,* Art. 7.

54 Ibid., Art. 6(2).

55 That formulation stayed also in the final text, cf. *Ideas Specific Programme*, 260.

56 See the report of the European Court of Auditors, specifically the quote already cited in Chapter 1, 'Financial Year 2005', 134.

57 'Special Report on the Management of Indirect RTD Actions under the Fifth Framework Programme (FP5) for Research and Technological Development (1998 to 2002), Together with the Commission's Replies' (Luxembourg: European Court of Auditors, 2004); François Colling, 'Management of Indirect RTD Actions under the 5th Framework Programme (FP5) for Research and Technological Development (1998 to 2002). Presentation to Budgetary Control Committee (COCOBU), European Parliament' (Strasbourg: European Court of Auditors, 19 April 2004); 'Synthesis of the Commission's Management Achievements in 2006' (Brussels: European Commission, 30 May 2007), 3–4; Marc Bellens, 'The Ex-Post Audits: The Example of FP6', Presentation (Brussels: European Commission, 2 June 2008).

58 Antonis Ellinas and Ezra Suleiman, *The European Commission and Bureaucratic Autonomy: Europe's Custodians* (Cambridge: Cambridge University Press, 2012), 58; similarly, Antonis Ellinas and Ezra Suleiman, 'Reforming the Commission: Between Modernization and Bureaucratization', *Journal of European Public Policy* 15, no. 5 (2008): 708–25; Michael W. Bauer, 'Impact of Administrative Reform of the European Commission: Results from a Survey of Heads of Unit in Policy-Making Directorates', *International Review of Administrative Sciences* 75, no. 3 (2009): 459–72.

59 'Spaniard Takes over the Reins at DG Research', Press Release (Cordis: 25005, 5 January 2006).

60 Cf. 'Achievements in Agricultural Policy under Commissioner Franz Fischler (Period 1995–2004)' (Brussels: European Commission, 2004).

61 William Cannell, 'The Ideas Programme. Objectives and Organization', Presentation (Brussels: European Commission, 11 June 2006), 5 [TK]. The notion 'Wachhund' (watchdog) has been used by Martin Bohle in his interview (see Appendix 2).

5 STATE OF CRISIS

1 Angela Merkel, 'Welcome and Opening Remarks' (Berlin: ERC ScC Plenary #09, 27 February 2007), 10.
2 Cit. in 'Report on ERC Launch Event Berlin' (Barcelona: ScC Extended Chairs #02, 17 April 2007), 2.
3 Ibid., 7.
4 *NSF Act*; Lewis and Selin, *Sourcebook*.
5 See also in König, 'Mission Accomplished?', 128.
6 For a full overview, see Figure 4.3.
7 Following Framework Programme regulations, the Specific Programme had to be formally supervised by a committee with representatives from each member state; cf. Gornitzka and Metz, 'Dynamics', 98.
8 Flink, 'Frontier Research.'
9 Gornitzka and Metz, 'Dynamics', 100.
10 *Ideas Specific Programme*, 260.
11 The distinction between agency model and fiduciary principle follows Giandomenico Majone, 'Two Logics of Delegation: Agency and Fiduciary Relations in EU Governance', *European Union Politics* 2, no. 1 (2001): 103; the principal-agency concept is popular among social scientists, for it provides a simplified template to problematize delegation (be it at individual level, in the private sector, or in the public sphere); cf. Joseph E. Stiglitz, 'Principal and Agent', ed. Steven N. Durlauf and Lawrence E. Blume, *The New Palgrave Dictionary of Economics* (Palgrave Macmillan, 2008); the model of an 'intermediary organization' is also based on the principle-agency theory. According to this interpretation of the theory, funding agencies, such as the NSF, are 'agents'. In the light of Majone's distinction, however, this appears to be overstretching the principal-agency-concept. It would be more appropriate to call the NSF (and other independent funding agencies) a fiduciary of two principals.
12 Natasha Gilbert, 'The Labours of Fotis Kafatos', *Nature* 464, no. 7285 (1 March 2010): 20–20; Colin Macilwain, 'Fork in the Road. Will the New European Research Council Lead EU Science to Success or Lose Its Way?', *The Scientist*, 1 February 2010.
13 Nowotny, 'Preserve'; I have argued similarly in König, 'Mission Accomplished?', 131, that Scientific Council leadership and ERCEA management had established a certain socio-organizational fabric.
14 'Report Formal Establishment', 2.
15 'ERC Board Meeting, 17.7.2007' (Tallinn: ScC Extended Chairs #03, 5 September 2007), 1; the particular issue was quickly resolved, as ' [t]he necessary legal arrangements have been made to reimburse the travel costs of the Principal Investigators coming to Brussels for the second stage interviews.' '3rd ERC Extended Chairs Meeting, Tallinn, 5.-6.9.2007' (Dublin: ERC ScC Plenary #12, 8 October 2007), 1; in the meantime, the Scientific Council

leadership had brought the case also to Commission President Barroso, but with little support, cf. 'Informal Exchange of Views between the ERC Scientific Council and Commission President Barroso, in Presence of Minister Gago (EU-Presidency)' (Tallinn: ERC ScC Extended Chairs #03, 5 September 2007).

16 Some emerged due to different expectations with regards to timing. For example, to solve the provision that each remote reviewer had to mail a blue-ink signature before he or she was allowed to review proposals (even though there was no payment involved), Jack Metthey's team put together proposals for simplifying procedures. As Commission protocol demanded, those proposals had to be examined by various legal and budget departments within the Commission services. Another issue was that, sometimes, the administrative branch was not responsible for obstacles. For example, the time to grant after the funding decision turned out to be surprisingly lengthy – but not, as the Scientific Council had suspected initially, because of the high 'administrative requirements' demanded by the Commission, but because of various delays due to communication mishaps between the PI and the host institution; cf. '16th ERC Scientific Council Meeting, Berne, 2.-3.7.2008' Minutes (Vilnius: ERC ScC Plenary #17, 10 September 2008), 3.

17 *ERC Decision*, Art. 7.

18 'Minutes 6th Meeting, 5.-6.10.2006', 3.

19 Babis Savakis, 'Quo Vadis ERC' (Vilnius: ERC ScC Plenary #17, 10 September 2008), 4; Ernst-Ludwig Winnacker, 'Some Thoughts on a Future Structure of the ERC' (Les Treilles: ERC ScC Retreat, 16 October 2008), 4.

20 Winnacker has told the trifles around his term as Secretary General in an autobiographical account. Cf. Winnacker, *Aufbruch*.

21 As Winnacker criticized in a Board meeting once, 'there is the need to adopt a formal decision in order for the Secretary General to formally participate in ERC DIS-Management and Board meetings'; 'ERC Board Meeting, 2.4.2009' (Brussels: ERC ScC Plenary #20, 28 April 2009); similarly, he complained later that implementing the position of the secretary general had not been 'accompanied by granting formal responsibilities, administrative or financial [. . .]. Thus, he was neither able nor permitted to participate in staff- or management meetings or make administrative decisions. Fortunately, the Director of the DIS was kind enough to organize weekly management meetings just to keep the Secretary General informed of major developments. In addition, the Secretary General was permitted to participate intensively in the preparation of the various calls-for-proposals and in the refinement of the peer-review system.' Ernst-Ludwig Winnacker, 'Mid-Term Review: Mission Statement ERC' (Brussels: ERC ScC Plenary #18, 18 December 2008), 6.

22 The outcry was reflected in the minutes: 'several voices: that the date (numbers/names etc.) should be available to ScC members first!!' Cf.'12th ERC Scientific Council Meeting, 8.-9.10.2007', Informal Minutes (Dublin: ERC ScC October 2007), 1–2 [TK].

23 '12th Meeting of the ERC Scientific Council, Dublin, 8.-9.10.2007', Minutes (Stockholm: ERC ScC Plenary #13, 17 December 2007), 1.

24 One time it was about the names of applicants submitting proposals, and later about the nomination of new members to the Scientific Council.

25 *ERC Decision*, Art 4(q).

26 José Manuel Silva Rodríguez, 'Provision of Administrative Support' (Brussels: ERC ScC Plenary #07, 10 January 2007).

27 'ERC Board Meeting, 17.1.2007', Minutes (Brussels: ERC ScC Plenary #09, 26 February 2007), 3: 'The President notes with regret the developments that led to the current absence of a provision under FP7 for staff to assist the Chair and Vice Chairs of the Scientific Council after 30 June 2007. The current experienced staff, selected by the Chairs and hosted by their home institutions so that we have immediate access to them, have been and are essential for the Chairs to fulfil their tasks, and therefore for the Scientific Council to function properly. We understand that alternative modes for financing this essential support are being explored, and of course we appreciate the importance of appropriate FP7 procedures. Equally, we would be remiss not to point out that absence of this essential support would (for the first time) de facto compromise the autonomy of the ERC Scientific Council and would seriously undermine the success of the ERC.'

28 Jack Metthey, 'Support to the Chair and the Vice-Chairs of the ERC Scientific Council' (Brussels: ERC ScC Plenary #07, 10 January 2007).

29 The support line is also briefly described in the Introduction to this book as the first of the author's four routines.

30 Why the Commission was making the issue so complicated is difficult to find out. Maybe it was a way of flexing its muscle in front of Kafatos and his peers; maybe it was also part of an internal power struggle within DG Research ranks; or maybe it was indeed just that the officers in charge were obsessively correct. In any case, the Scientific Council leadership felt so constrained that, at least once, Nowotny (and, following her, also Kafatos) threatened to resign and to go public. Wondrously, the Commission found a solution in the short term; cf. '18th ERC Scientific Council Meeting, Brussels, 18.-19.12.2008', Informal Minutes (Brussels: ERC ScC, December 2008), 1 [TK]. One year later, an audit alleged that a substantial part of the initial funding had not been in line with the financial regulations. Specifically in the case of Kafatos' host institution, this concerned costs for work that was done before the contract was signed, and costs for work exceeding the official time frame, while – in the case of Nowotny's institution – it concerned costs for office space and maintenance; cf. 'ERCSC-CHAIR-SUP 044505', Audit Report (Brussels: European Commission, 2 March 2009) [TK]. As a consequence, both Kafatos' and Nowotny's host institutions had to repay a substantial five-digit sum.

31 Fotis C. Kafatos, 'Talking Points to Mid-Term Review Panel Meeting' (Brussels: ERC ScC Plenary #20, 28 April 2009), 1–2.

32 The Commissioner also warned his colleagues that the 'consequences of not getting them up and running in time would be close to disaster'. Janez Potočnik, 'Speaking Points on Executive Agencies', Unpublished Document (Brussels: European Commission, 29 August 2007), 2, 4 [TK].

33 *Executive Agencies*, (4).

34 'Delegating Implementing Tasks to Executive Agencies', 14, 19.

35 *Decision Setting up the 'European Research Council Executive Agency' for the Management of the Specific Community Programme 'Ideas' in the Field of Frontier Research in Application of Council Regulation (EC) No 58/2003* (European Commission: EC/2008/37 [O.J. L 9/15], 14 December 2007), Art. 5.

36 Fotis C. Kafatos, 'In Strict Confidence', (Athens: ERC ScC Plenary #14, 20 February 2008).
37 'Meeting between President Barroso and Representatives of the Scientific Council and Secretary General of the ERC' (Brussels: ERC ScC Plenary #15, 20 April 2008), 3. It must have been a memorable though utterly frustrating meeting for Kafatos and his company, Nowotny, Estève, and Winnacker. After Kafatos made his case, Barroso politely expressed 'his support to ensure maximum flexibility in addressing the identified challenges', with the explicit (and unsurprising) restriction that this 'can only be exercised within the framework of rules'. He then delegated the affair back to Silva Rodríguez by asking him 'to indicate in which direction the solutions would go'; the Director General vaguely promised 'intermediate solutions would continue to be explored'.
38 '14th ERC Scientific Council Plenary Meeting', Agenda (Athens: ERC ScC Plenary #14, 28 February 2008), 1.
39 'Non-Paper Outline for Discussion with Janez Potočnik' (Brussels: ERC ScC Plenary #16, 14 May 2008), 2.
40 Fotis C. Kafatos, 'Memo to the Competitive Council Concerning Difficulties for the ERC in Its Transition Process towards an Executive Agency', Unpublished Document (Brussels, 1 July 2008), 1 [TK].
41 Fotis C. Kafatos, '[Attachment to Letter to Scientific Council]', Annex to Minutes (London, February 20, 2008), 15.
42 '13th ERC Scientific Council Meeting, Stockholm, 17.-18.12.2007', Minutes (Athens: ERC ScC Plenary #14, 28 February 2008), 1.
43 Winnacker voiced concerns for the first time in early 2008, cf. 'Preparatory Meeting on the Establishment and Operations of the ERCEA Steering Committee, 14.2.2008' (Brussels: ERC ScC Plenary #15, 29 April 2008), 2–3; Ernst-Ludwig Winnacker, 'Problems Associated with the Creation of an Executive Agency' (Brussels: ERC ScC Plenary #14, 10 February 2008).
44 '11th Meeting of the ERC Scientific Council, Lisbon, 3.-4.7.2007', Minutes (Tallinn: ScC Extended Chairs #03, 5 September 2007), 2.
45 Janez Potočnik, 'Today's Debate on Agencies' (Brussels: ScC Extended Chairs #02, 5 September 2007).
46 In an early presentation, Cannell stated that the agency 'is expected to be fully operational by mid-2008', cf. Cannell, 'Ideas', 16; in the autumn of 2007, Jack Metthey reassured the Scientific Council that '[t]he agency is foreseen to be operational in summer 2008, although this might be slightly optimistic.' 'Minutes 3rd Extended Chairs Meeting, 5.-6.9.2007', 1.
47 '15th ERC Scientific Council Meeting, Brussels, 29.-30.4.2008', Informal Minutes (Brussels: ERC ScC, April 2008), 1 [TK].
48 '17th ERC Scientific Council Meeting, Vilnius, 10.-11.9.2008', Informal Minutes (Vilnius: ERC ScC, 10 September 2008), 2 [TK].
49 Jack Metthey, 'ERCEA State of Play', Presentation (Vilnius: ERC, 10 September 2008), ScC Plenary #17.
50 '18th ERC Scientific Council Meeting, Brussels, 18.-19.12.2008', Annotated Minutes (Istanbul: ERC ScC Plenary #19, 11 March 2009), 2; cf. also Jack Metthey, 'ERCEA State of Play' (Brussels: ERC ScC Plenary #20, 28 April 2009); Jack Metthey, 'ERCEA State of Play' (Brussels: ERC ScC Plenary #21, 30 June 2009).
51 At the time, it was already well known within Commission services how long

it took until an executive agency would be operational: a report had concluded in 2005 'that the time needed to plan and set up an executive agency should not be underestimated. The first agencies established point to a lead time of about two years from the Commission decision creating the agency to the moment when the agency is fully operational'; quoted in 'Delegating Implementing Tasks to Executive Agencies', 16.

52 'Non-Paper', 2.

53 Winnacker, 'Problems.'

54 Kafatos' preparatory remarks to the meeting read like a list of strong demands: 'Appoint Jack Metthey immediately as Director of the ERC-Executive Agency'; 'Permit the ERC-EA Director to advertise outside the Commission for Heads and Deputy Heads of Units wherever required'; 'Permit him to hire the necessary and suitable personnel'; 'Accelerate the Delegation Act and the official Appointment of the Steering Committee'; cf. Fotis C. Kafatos, 'Meeting with President Barroso' (Brussels: ERC ScC Plenary #15, 16 March 2008), 2.

55 For various reasons, such comparisons have to be taken with a grain of salt. The numbers can be found in 'Delegating Implementing Tasks to Executive Agencies', 16.

56 'Two New Executive Agencies to Manage European Research', Press Release (Cordis: 28853, 14 December 2007); 'EU's Research Executive Agency Becomes Autonomous', Press Release (Cordis: 30903, 15 June 2009).

57 Kafatos, 'Competitive Council', 1.

58 The quote was originally from a 'former senior official of the Commission Legal service'; cf. 'ERC Review of Structures and Mechanisms' (Brussels: European Commission, 27 March 2009), 5 [TK]; see also 'Review of the ERC Structures and Mechanisms: Position Paper of the ScC' (Brussels: ERC Scientific Council, 26 March 2009) [TK].

59 *Ideas Specific Programme*, 266–7; cf. also Chapter 3.

60 *Communication to the Council and the European Parliament on the Methodology and Terms of Reference to Be Used for the Review to Be Carried out by Independent Experts Concerning the European Research Council Structures and Mechanisms* (European Commission: COM/2008/526, 26 August 2008), 3.

61 Savakis, 'Quo Vadis ERC', 2; 'Position Paper', 13–14.

62 Ernst-Ludwig Winnacker, 'Midterm Review – Draft Mission Statement ERC Scientific Council' (Brussels: ERC ScC, 10 February 2009) [TK].

63 Mathias Dewatripont, '[to Scientific Council]', email draft (5 December 2008) [TK]. I am grateful to Mathias Dewatripont for letting me quote from this email.

64 'Position Paper', 5, 7, 8–9.

65 Janez Potočnik, 'Statement for the VIP Corner on the European Research Council', Unpublished Document (Brussels: European Commission, 23 July 2009), 1 [TK]. The other members of the panel were David Sainsbury who (as noted in Chapter 3) had played a vital role in the UK's support for the ERC; economist Fiorella Kostoris from La Sapienza University in Rome; Lars-Hendrik Röller, former chief economist of the European Commission; and Elias Zerhouni, former director of the National Institutes of Health in the US.

66 Martin Enserink, 'Fix Funding Agency's "Original Sin," ERC Review Panel Demands', *Science* 325, no. 5940 (31 July 2009): 523; similarly, Natasha

Gilbert, 'European Body Told to Cut Free', *Nature* 460, no. 7255 (24 July 2009): 557.

67 Quotes in this and the next paragraph are from Vaira Vīķe-Freiberga et al., 'Towards a World Class' Frontier Research Organisation. Review of the European Research Council's Structures and Mechanisms', Midterm Report (Brussels: European Commission, 23 July 2009).

68 Ibid., 23.

69 This was probably reinforced by a dry yet sobering legal opinion from a sympathetic director in the Commission's legal service, cf. Jürgen Grunwald, 'Review of Structures and Mechanisms of the European Research Council (ERC)', Unpublished Document (Brussels: European Commission, 12 June 2009) [TK].

70 Vīķe-Freiberga et al., 'Towards', vi.

71 Ibid., 27.

72 *Communication to the Council and the European Parliament: The European Research Council – Meeting the Challenge of World Class Excellence* (European Commission: COM/2009/552, 22 October 2009), 4.

73 Helga Nowotny, 'Confidential Memo of the ScC Closed Session, 18.10.2009' (Brussels: ERC ScC Plenary #22, 13 October 2009), 1.

74 'Position Paper', 9.

75 Nowotny, 'Confidential Memo of the ScC Closed Session, 18.10.2009', 1.

76 Kafatos, 'Towards an Integrated ERC', 2.

77 Winnacker, 'Midterm Review – Draft Mission Statement ERC Scientific Council.'

78 Mas-Colell had been already appointed in 2006, and Nowotny had shared responsibility for a while previously – upon Kafatos' request, she had been chairing the Scientific Council meetings in 2009; cf. '23rd Scientific Council Meeting, 15 – 16.12.2009', Annotated Minutes (Bucharest: ERC ScC Plenary #24, 11 March 2010), 8.

79 Helga Nowotny, 'Follow-up MTR Input [to Fotis Kafatos, Daniel Estève, Ernst-Ludwig Winnacker, Jack Metthey, William Cannell]', email (4 May 2009) [TK]. I am grateful to Helga Nowotny for letting me quote from this email.

80 Helga Nowotny, '[to Máire Geoghegan-Quinn]', (Brussels: ERC ScC Plenary #27, 30 July 2010).

81 *Publication of a Vacancy for a Post of Principal Adviser (grade AD 14) — Director Designate of the European Research Council Executive Agency (ERCEA) in Brussels (Article 29(2) of the Staff Regulations)* (European Commission: COM/2009/10222 [O.J. C 295 A/1], 20 March 2006); 'Post of Principal Adviser (Grade AD 14) Director Designate of the European Research Council Executive Agency (ERCEA)', Pre-selection Panel Report (Brussels: European Commission, 10 June 2010) [TK].

82 Nowotny, 'Confidential Memo of the ScC Closed Session, 18.10.2009', 1.

83 Nowotny, ' [to Máire Geoghegan-Quinn]'; the internal Commission guidelines for appointing senior officials span nearly 50 pages: 'Guidelines for Commission Services on Appointment Procedures for Senior Officials' (Brussels: European Commission, October 2008) [TK].

84 Nowotny, '[to Máire Geoghegan-Quinn].'

85 Ramon Marimon has published his correspondence on the issue on his website, documenting how he went on writing letters of complaint to the

Commissioner, to Robert-Jan Smits, to the Scientific Council, and to the European Ombudsman – he insisted that 'what happened is not so much bad news for me, but for the ERC'; see Ramon Marimon, '[letter to ERC Scientific Council]' (5 November 2010) [RMS]. The other two contenders were Donald Dingwell, ERC Grantee and later the third Secretary General of the ERC, albeit with a much reduced influence; and Carl-Henrik Heldin who, for the time of the selection process, had temporarily stepped down from his membership of the Scientific Council, and who, between 2011 and 2014, would act as one of its Vice Chairs.

86 'Vacancy Notice for the Post of "Principal Adviser in Directorate-General Research, Director Designate of the European Research Council Executive Agency (ERCEA) in Brussels" [to Ramon Marimon Sunyol]' (29 November 2010) [RMS].

87 Nowotny, '[to Máire Geoghegan-Quinn].'

88 'European Commission Has Set up Task Force to Help Maximise the Potential of the European Research Council', Press Release (Rapid: IP/10/1759, 21 December 2010).

89 Robert-Jan Smits et al., 'European Research Council Task Force', Final Report (Brussels: European Commission, 2011), 2 [TK].

90 Ibid., 8.

91 A comprehensive overview was provided in 'ERC Task Force Recommendations Follow up' (Brussels: ERC ScC Plenary #32, 20 November 2011).

92 The decision was already made before summer 2013; however, it took the Commission until December to make the announcement. By then, all relevant news outlets had already featured the 'rumours'. Cf. 'Jean-Pierre Bourguignon Appointed next President of the European Research Council', Press Release (Rapid: IP/13/1260, 17 December 2013).

93 When proposing the next MFF 2014–2020, the Commission not only took pre-emptive measures by producing evidence of its own frugality, but also inc luded a series of measures of 'simplification', consisting of two types of reform: re-packaging the programmes run by the Commission (for example, putting several independent programmes under the 'Horizon 2020' framework), and reducing administrative costs – basically saying that it would continue outsourcing the conduct of its programmes to executive agencies more aggressively. *A Budget for Europe 2020* (European Commission: COM/2011/500 [O.J. C 264/18], 29 June 2011), 21–23. Thus, the Commission stipulated more streamlining of 'common support' services between different executive agencies, and attempted to bring the agency into line along a series of administrative, staff, and financial procedures.

94 Martin Bohle, 'The Common Support Centre', Presentation (Brussels: EARTO, 4 February 2014) [TK]; Theodore Papazoglou and Maria Olivan Aviles, 'Current State of Play', (Utrecht: ERC ScC Plenary #42, 22 October 2013).

6 A RATHER CONVENTIONAL SYSTEM

1 'The ESII–2007 Call. A Scenario' (Vienna: ERC ScC Plenary #03, 26 April 2006), 2; 'Panel Recruitment: Status and next Steps' (Berlin: ERC ScC Plenary #09, 26 February 2007), 6.
2 'Impressive Demand for First European Research Council Grants', Press Release, (Brussels: ERC ScC, 26 April 2007) [TK]; also 'Starting Grant First Results', Presentation (Prague: ERC ScC Plenary #10, 14 May 2007).
3 'First Conclusions of StG1' (Prague: ERC ScC Plenary #10, 14 May 2007), 6.
4 'Starting Grant First Results', 8–10; 'Update on Implementation of Plan B', (Prague: ERC ScC Plenary #10, 14 May 2007), 2; '10th Meeting of the ERC Scientific Council, Prague, 14.–15.5.2007', Minutes (Lisbon: ERC ScC Plenary #11, 3 July 2007), 1.
5 'The Scientific Council stressed that each proposal that is discussed in the panel meetings should have been evaluated by a reviewer that is present at the meeting. [. . .] The Scientific Council also stressed the importance of briefing the panels on the review procedure for interdisciplinary proposals?. Cf. 'Minutes 10th Meeting, 14.–15.5.2007', 1–2.
6 Pavitt, 'Radical Proposal.'
7 EURAB, 'European Research Council', 1.
8 Gannon, 'Summary'; van Dyck, 'Follow-Up'; van Dyck, 'ERC for All Sciences'; van Dyck, 'ERC.'
9 As Mitsos had explained, 'when we move towards grants, then bureaucracy can be avoided'. Cf. Mitsos, 'Closing Address', 5.
10 Daniel Estève, 'Proposal for Discussion: Call for Grants to Young Investigators and Newly Established Teams' (Brussels: ERC ScC Plenary #02, 24 January 2006).
11 Helga Nowotny, 'The ERC: Some Preliminary Ideas for Further Discussion on What to Do and How to Proceed' (Brussels: ERC ScC Plenary #02, 30 December 2005).
12 '2nd Meeting of the ERC Scientific Council, 24.–25.1.2006', Minutes (Vienna: ERC ScC Plenary #03, 26 April 2006), 2, [PE].
13 William Cannell, 'ERC Launch Strategy' (Vienna: ERC ScC Plenary #03, 10 April 2006), 1.
14 Nowotny, 'Preliminary Ideas.'
15 William Cannell, 'The Early Stage Independent Investigators Scheme (ESII)' (Vienna: ERC ScC Plenary #03, 10 April 2006), 3.
16 Estève, 'Proposal', 1.
17 Cannell, 'ESII', 2.
18 'It was agreed that a more appealing name should be sought for the EI-scheme'. Cf. 'Minutes Extended Chairs' Meeting'; and 'Minutes 3rd Meeting, 26.–27.4.2006', 2.
19 Cannell, 'ERC Launch Strategy', 2.
20 *ERC Work Programme 2013* (European Commission: C/2012/4562, 9 July 2012).
21 Andreu Mas-Colell et al., 'Food for Thought and Annexes' (Rehovot: ERC ScC Plenary #23, 15 December 2009), 8.
22 '23rd ERC Scientific Council Meeting, Rehovot 15.–16.12.2009', Informal Minutes (Rehovot: ERC ScC December 2009), 16–19'[TK].
23 It should be mentioned that, up until the introduction of the new scheme, the

'Advanced Grant' scheme had foreseen that, in order to 'encourage interdisciplinarity', a Principal Investigator could name a 'co-investigator'; cf. *ERC Work Programme 2008* (European Commission: C/2007/5746, 29 November 2007), 4. However, this was not made use of often, probably because it was explicitly called an exception' and came with additional obligations in the decision-making procedure.

24 'The ScC decided by a vote (15 votes in favour and 3 abstentions) that the new scheme will be introduced in the WP2012. The ScC will provide written comments to it in January 2011 and have a final discussion on it at its next plenary meeting in Brussels in February 2011.' '28th Scientific Council Plenary Meeting, Brussels, 16.–17.12.2010', Annotated Minutes (Brussels: ERC ScC Plenary #29, 8 February 2011), 3.

25 'Synergy Grant Final Results', (Brussels: ERC ScC Plenary #38, 5 December 2012), 17.

26 For an overview, see Stephan, *How Economics Shapes Science*, 129–141.

27 For a general, and very useful overview of categories in decision–making on allocation of scarce goods, see Jon Elster, *Local Justice: How Institutions Allocate Scarce Goods and Necessary Burdens* (New York: Russell Sage Foundation, 1992), 62–112; there are a few noteworthy reflections on the role of peer review in research funding; cf. Daryl E. Chubin and Edward J. Hackett, *Peerless Science: Peer Review and U. S. Science Policy* (Albany: SUNY Press, 1990), 2–20; John Ziman, 'What Are the Options? Social Determinants of Personal Research Plans', *Minerva* 19, no. 1 (March 1981): 1–42; Arie Rip, 'The Republic of Science in the 1990s', *Higher Education* 28, no. 1 (1994): 3–23; James McCullough, 'The Role and Influence of the US National Science Foundation's Program Officers in Reviewing and Awarding Grants', *Higher Education* 28, no. 1 (1994): 85–94; Arie Rip, 'Higher Forms of Nonsense', *European Review* 8, no. 4 (2000): 467–85; Stefan Hirschauer, 'Peer Review Verfahren auf dem Prüfstand', *Zeitschrift Für Soziologie* 33, no. 1 (2004): 62–83; Liv Langfeldt, 'The Policy Challenges of Peer Review: Managing Bias, Conflict of Interests and Interdisciplinary Assessments', *Research Evaluation* 15, no. 1 (2006): 31–41; Martin Reinhart, *Soziologie und Epistemologie des Peer Review* (Baden-Baden: Nomos, 2012).

28 Stephan Lorenz, 'Procedurality as Methodological Paradigm. Or: Methods as Procedures', *Forum: Qualitative Social Research* 11, no. 1 (26 November 2009): 8; the first to analyse the distinct function of procedures in social systems was Niklas Luhmann, *Legitimation durch Verfahren* (Frankfurt a. M.: Suhrkamp, 1969).

29 For an overview of the early decades, see Stephen Cole, Leonard Rubin, and Jonathan R. Cole, *Peer Review in the National Science Foundation: Phase One of a Study: Prepared for the Committee on Science and Public Policy of the National Academy of Sciences* (Washington, D.C.: National Academy Press, 1978), 11–16.

30 Cf. Philippe Laredo and Philippe Mustar, eds., *Research and Innovation Policies in the New Global Economy: An International Comparative Analysis* (Cheltenham: Edward Elgar Publishing, 2001); for the US, see Sheila Slaughter and Gary Rhoades, 'From "Endless Frontier" to "Basic Science for Use": Social Contracts between Science and Society', *Science, Technology & Human Values* 30, no. 4 (2005): 536–72; for Europe

specifically, see also Inga Ulnicane, 'Broadening Aims and Building Support in Science, Technology and Innovation Policy: The Case of the European Research Area', *Journal of Contemporary European Research* 11, no. 1 (2015): 31–49.

31 Up until the 1990s, the NSF, for example, had simply asked reviewers to assess proposals 'in terms of their specific technical and scientific merit.' Robert Frodeman and Jonathan Parker, 'Intellectual Merit and Broader Impact: The National Science Foundation's Broader Impacts Criterion and the Question of Peer Review', *Social Epistemology* 23, no. 3–4 (2009): 337–45. The agency was then forced by US Congress to explicitly ask for 'the broader impacts of a project on society' as well; cf. J. Britt Holbrook and Robert Frodeman, 'Science: For Science's or Society's Sake?', *Science Progress*, 1 March 2012. In 2011, the NSF paid tribute by also requiring a separate account of broader impacts in the final project report. Proposals from now on would not only be assessed upon their scientific quality, but also in regards to the question of if they would yield societal, economic benefits; cf. Julia R. Kamenetzky, 'Opportunities for Impact: Statistical Analysis of the National Science Foundation's Broader Impacts Criterion', *Science and Public Policy* 40, no. 1 (2013): 72–84.

32 Whatever the 'mixture of motives' may be, it occasionally leads 'to conflicts of purpose and to behavioral consequences in which the acquisition of verified information does not serve as highest maxim, but recedes before other ideals.' Fraud, conflicts of interest, and plagiarism are all aspects of 'scientific malfunctions'; inevitably, the peer-reviewed allocation of funding, where scientists judge over their peers and award a dually valuable grant, is one of its trouble spots. Klaus Fischer, 'Science and Its Malfunctions', *Human Architecture: Journal of the Sociology of Self-Knowledge* 6, no. 2 (2008): 2, mentions worldly motives such as 'to be famous or rich, to attain social recognition or authority, to gain power over others or over nature, to increase utility for themselves or society, to find peace of mind, to reach certitude of salvation, to pursue religious goals, to further political ideologies, or to save the world.'

33 There is little evidence how wide-spread scientific malpractice actually is, but the informed guess is that one out of ten research results is fabricated; cf. Daniele Fanelli, 'How Many Scientists Fabricate and Falsify Research? A Systematic Review and Meta-Analysis of Survey Data', ed. Tom Tregenza, *PLoS ONE* 4, no. 5 (2009): e5738.

34 In 1997, a now famous study claimed that 'nepotism and sexism' lay at the heart of the Swedish funding agency's peer review procedure. Christine Wennerås and Agnes Wold, 'Nepotism and Sexism in Peer-Review', *Nature* 387, no. 6631 (1997): 341–43.

35 Danielle L. Herbert et al., 'On the Time Spent Preparing Grant Proposals: An Observational Study of Australian Researchers', *BMJ Open* 3, no. 5 (2013): e002800.

36 Rip, 'Nonsense', 468.

37 The quote is from a survey from the 1980s on 'attitudes towards peer review' among US researchers; in the survey, almost 2/3 of the participants in the survey supported the statement that 'Reviewers are reluctant to support unorthodox or high-risk research'; cf. Chubin and Hackett, *Peerless Science*, 66; according to a recent survey of existing literature on the topic, 'the

majority of the research on peer review concludes that it is inherently conservative and unable to select truly innovative research proposals'. Cf. Terttu Luukkonen, 'Conservatism and Risk-Taking in Peer Review: Emerging ERC Practices', *Research Evaluation* 21, no. 1 (2012): 48–60.

38 To refer just to some recent contributions in different contexts, cf. Richard Smith, 'Peer Review: A Flawed Process at the Heart of Science and Journals', *Journal of the Royal Society of Medicine* 99, no. 4 (2006): 178–82; Paul Nightingale and Alister Scott, 'Peer Review and the Relevance Gap: Ten Suggestions for Policy-Makers', *Science and Public Policy* 34, no. 8 (2007): 543–53; Joshua M. Nicholson and John P. A. Ioannidis, 'Research Grants: Conform and Be Funded', *Nature* 492, no. 7427 (6 December 2012): 34–6; David Gurwitz, Elena Milanesi, and Thomas König, 'Grant Application Review: The Case of Transparency', *PLoS Biol* 12, no. 12 (2014): e1002010.

39 Fotis C. Kafatos, 'Cover Letter on Principles Evaluation Panel Structure', (Helsinki: ERC ScC Plenary #05, 22 June 2006).

40 'Extended Chairs' Meeting', 3.

41 'ERC Consolidated Panel Structure and Titles', (London: ERC ScC Plenary #06, 10 May 2006).

42 'The ESII–2007 Call. Possible Methodology for Establishing the Panels and Referee Base. Discussion Paper' (Vienna: ERC ScC Plenary #03, 26 April 2006); 'The 2007 SG Evaluation. Options: Timing, Interview'(Copenhagen: ERC ScC Plenary #04, 29 May 2006); 'Evaluation Structure for the StG–1 Call. Information and Points of Discussion for the Scientific Council' (Helsinki: ERC ScC Plenary #05, 22 June 2006). Some assumptions made by the Scientific Council and the Secretariat at this early stage (such as, that it would be possible to draft a full proposal after the first evaluation round; and that the final decision should be taken by the Scientific Council) would later be rescinded, because they did not conform either with FP7 provisions or with the logic of the procedure.

43 *ERC Rules for the Submission of Proposals and the Related Evaluation, Selection and Award Procedures for Indirect Actions under the Ideas Specific Programme of the Seventh Framework Programme* (2007–2013) (European Commission: C/2007/2286 6 June 2007), 5; and *Amending Decision C(2007)2286 on the Adoption of ERC Rules for the Submission of Proposals and the Related Evaluation, Selection and Award Procedures for Indirect Actions under the Ideas Specific Programme of the Seventh Framework Programme (2007 to 2013)* (European Commission: C/2010/767 [O.J. L 327/51], 9 December 2010), 53.

44 'Evaluation Criteria for Starting Grant' (Copenhagen: ERC ScC Plenary #04, 29 May 2006), 3.

45 'ERC Work Programme 2007', Draft (Helsinki: ERC ScC Plenary #05, 3 July 2006), 8.

46 '5th Meeting of the ERC Scientific Council, 3.–4.7.2006', Minutes (London: ERC ScC Plenary #06, 5 October 2006), 3.

47 *ERC Work Programme 2007* (European Commission: C/2006/561, 26 February 2007).

48 Pavel Exner, Carl-Henrik Heldin, and Carlos Duarte, 'Reports ScC Delegates to Various Panel Meetings' Observation report (Rehovot: ERC ScC Plenary #23, 15 December 2009), 2.

49 The ERC also ran a public consultation, which was not met with great

enthusiasm; cf. Alejandro Martin-Hobdey, 'Points to Raise / Clarify during Panel Structure Discussion and during Panel and Referee Recruitment Discussion', Briefing Document (Brussels: ERCEA, 2 July 2006), 2 [AMH]; 'Consultation on Panel Members and Panel Chairs: State of Play' (Helsinki: ERC ScC Plenary #05, 22 June 2006).

50 'The ESII–2007 Call. Possible Methodology.'

51 'Proposed Methodology for the Choice of Panel Members' (Helsinki: ERC ScC Plenary #05, 22 June 2006).

52 '4th Meeting of the ERC Scientific Council, 29.–30.5.2006', Minutes (Helsinki: ERC ScC Plenary #05, 3 July 2006), 3.

53 The Scientific Council's Rules of Procedure would also state that its members should not 'influence, under any circumstances' the peer review procedure, cf. 'Rules of Procedure and Code of Conduct of the Scientific Council' Unpublished Document (Brussels: ERC ScC, 8 February 2011), 4 [TK]; keeping its members at arm's length from the peer review procedure was important in order to further substantiate the impartiality of the decision-making and to impose the same standard of rules that were implemented for the panels.

54 Ian Halliday, 'The Ideas Work Programme Evaluation Process', Observation report (Brussels: ERC ScC, 14 July 2008), 2 [TK].

55 Kafatos, 'Cover Letter.'

56 Manolis Antonoyiannakis and Fotis C. Kafatos, 'The European Research Council: A Revolutionary Excellence Initiative for Europe', *European Review* 17, no. 3–4 (2009): 514.

57 Ernst-Ludwig Winnacker, 'On Excellence through Competition', *European Educational Research Journal* 7, no. 2 (2008): 126.

58 Carl-Henrik Heldin, 'The European Research Council — a New Opportunity for European Science', *Nature Reviews Molecular Cell Biology* 9, no. 5 (2008): 418.

59 Nowotny, 'Preserve.'

60 *Ideas Specific Programme*, 245.

61 In early 2010, when the volcano Eyjafjallajökull erupted in Iceland and air travel was halted across Europe, various panel meetings of the then running 'Starting Grant' call had to be postponed; nonetheless, the funding call was brought to a result. Cf. Alejandro Martin-Hobdey, 'Status StG 3 Presentation' (Brussels: ERC ScC Plenary #25, 28 April 2010).

62 This was not only the opinion of independent observers, such as Halliday, 'The Ideas Work Programme Evaluation Process'; Riitta Mustonen, 'The Ideas Work Programme Evaluation Process', Observation report (Brussels: ERC ScC, 10 October 2008) [TK], but also the finding of the survey conducted on behalf of the Mid-Term Review committee, cf. Vīķe-Freiberga et al., 'Towards'. Also, the overwhelming majority of more than 80 chairs of ERC review panels answering my survey in 2014 (cf. Introduction) held that the overall system was fair and balanced. Even though they could hardly be called impartial in that respect, this group consisted of powerful opinion-makers within the academic tribes.

63 The notion of orchestration was initially used by Rip, 'Republic of Science', 18.

64 'Possible Implications for the ERC Amended Work Programme of the High Demand for StG' (Prague: ERC ScC Plenary #10, 14 May 2007), 1.

65 The administrative branch was always wary of the workload; it proposed measures to deter applicants in a way that the best would still be willing to apply and to limit the workload of the panels. For example, the agency suggested in one particular instance 'automatic assignment of Cs at step 1 to proposals which fulfil specific criteria based on an average low mark and small standard deviation; no panel comments for proposals with C; evaluation of the PI based on marks assigned to pre-fixed categories for both AdG and StG; eligibility checks only for projects passing to Step 2; remote referees needed besides the ones identified in the Step 1 panel meeting to be identified by SOs and confirmed by the Chairs.' Cf. '33rd ERC Scientific Council Meeting, Brussels, 6.–7.12.2011', Annotated Minutes (Brussels: ERC ScC Plenary #34, 28 February 2012), 3.

66 Nowotny, 'Preliminary Ideas', 4; since its first edition in 2007, every annual Work Programme of the ERC proudly paraded the uniform sentence that 'excellence is the sole criterion of evaluation.' Cf. *ERC Work Programme 2007*, 11; *ERC Work Programme 2015* (European Commission: C/2014/5008, 22 June 2014), 30.

67 For example, in the sixth edition of the Fp format, 'excellence' was one evaluation criterion (out of five) for assessing the different funding streams under the NEST programme; cf. *Work Programme for the Specific Programme for Research, Technological Development and Demonstration: 'Integrating and Strengthening the European Research Area'* (European Commission: n.n., 30 September 2002), 25–7; also, a funding instrument appropriately called 'Networks of Excellence' was set up to 'strengthen scientific and technological excellence on a particular research topic' to fund 'a joint training programme for researchers and other key staff'; cf. European Commission, 'Classification of the FP6 Instruments. Detailed Description' (Brussels: European Commission, October 2004), 9.

68 In public, Kafatos probably used the signature phrase for the first time in early 2007; cf. 'Launch Event', 2; similarly, 'ERC: Activities and Achievements in 2008', 15 and 30. The notion resonated very well, and was even used by the official Commission Communication reacting to the Mid-Term Report, cf. *Meeting the Challenge*, 5.

69 As Kafatos explained to the chairs of the first 'Starting Grant' call panels, 'we intend to monitor very closely the implementation of our strategy during the operation of the first call for proposals', Fotis C. Kafatos, 'Why Is European Integration Important for Scientific Innovation?', Unpublished Document(Brussels: ERC ScC, 24 March 2007), 2, 4 [TK].

70 It should be emphasized here again that the permission to draft and adopt the ERC's Work Programme was an extraordinary privilege of the Scientific Council, established in the main legal texts, *Ideas Specific Programme*, 254; and *ERC Decision*, Art. 3(2). For an overview of additional guidance documents drafted under the authority of the Scientific Council, see the fourth layer of documents in Figure 4.3.

71 Vīķe-Freiberga et al., 'Towards', v.

72 'She Figures 2012. Gender in Research and Innovation. Statistics and Indicators' (Brussels: European Commission, 2013), 88.

73 Helga Nowotny, 'Advancing Gender Balance in Science – from the Perspective of a Funding Agency', Presentation (Vienna: ERC ScC, 30 November 2011), 12–13 [TK].

74 'Gender Equality Plan 2007 – 2013' (Brussels: ERC ScC, 19 February 2011), 2–4.
75 For example, the NSF's Office of Inspector General has broad authority to deal with 'fraud, waste, and abuse within the NSF or by individuals that receive NSF funding'. Its activities include not only financial audits, but also investigations in alleged 'cases of research misconduct' based on whistle-blowers; the NSF can terminate grants, punish host institutions and PIs and even blame them publicly. Cf. 'Semiannual Report to Congress' (Washington, D.C.: NSF Office of Inspector General, March 2015), 2.
76 The European Ombudsman investigates complaints by European citizens against the European institutions (but not vice versa); the Court of Auditors is strictly focused on financial affairs (which usually take place after the funding decision).
77 'Scientific Misconduct Strategy' (Brussels: ERC ScC, 5 October 2012), 1; for an internal discussion of the restricted options for the ERC, cf. '2nd Meeting of the ERC ScC Standing Committee on Conflict of Interest, Riga, 19 October 2011', Minutes (Brussels: ERC ScC Plenary #34, 28 February 2012).
78 Helga Nowotny and Pavel Exner, 'Improving ERC Ethical Standards', *Science* 341, no. 6150 (6 September 2013): 1043.
79 'ERC Grant Schemes Guide for Peer Reviewers' (Brussels: ERC ScC, 10 September 2007): 8*[TK]*.
80 *ERC Rules for Submission*, 10, 26.
81 Ibid., 26–7.
82 'Conflict of Interest' (Brussels: ERC ScC COIME Meeting #1, 16 June 2010); '1st COIME Meeting, Santiago, 29.6.2010', Minutes (Luxembourg: ERC ScC Plenary #27, 28 October 2010), 3. The issue could be alleviated through an amendment of the legal provision, cf. *ERC Amended Rules for Submission*.
83 In an article on occasion of publishing the ESF's report on research integrity, Marja Makarow observed that, while 'everybody wants research that can be trusted' it had been 'surprisingly difficult to produce a document that can be widely agreed.' Her tentative answer was that 'perhaps [. . .] Europe is a mosaic of different traditions of academic discipline'. See Marja Makarow, 'Towards a Code for Science', *Research Europe*, 8 July 2010.
84 'Minutes 2nd COIME Meeting', 3.
85 'Evaluation Criteria.'
86 Benjamin Turner, 'Granting and Evaluation Issues' (Rehovot: ERC ScC Plenary #23, 15 December 2009), 11.
87 The issue will be discussed in Chapter 7.
88 'Minutes 23rd Meeting', 6.
89 Turner, 'Granting', 3.
90 *ERC Work Programme 2008*, 25–6.
91 'ERC Board Meeting, 16.6.2008', Minutes (Bern: ERC ScC Plenary #16, 2 July 2008), 3.
92 *ERC Work Programme 2008*, 11.
93 Quote from Carl-Henrik Heldin in Alain Peyraube, Carl-Henrik Heldin, and Hans-Joachim Freund, 'Reports of the ERC ScC Delegates Who Attended the ERC–2010-AdG Call Panel Meetings', Observation report (Brussels: ERC ScC Plenary #28, 16 December 2010), 3–4.
94 'Possible Options Regarding the Final Panel Chair Meeting and the Evaluation of Cross-Panel Interdisciplinary Proposals' (Brussels: ERC ScC Plenary #27,

28 October 2010); '27th Scientific Council Plenary Meeting, Luxembourg, 28.–29.10.2010', Annotated Minutes (Brussels: ERC ScC Plenary #28, 16 December 2010), 4. See also *ERC Work Programme 2012* (European Commission: C/2011/4961, 19 July 2011), 13–4.

95 'Evaluation of ID Proposals: Preliminary Analysis' (Brussels: ERC ScC Plenary #40, 18 March 2013), 37.

96 Fotis C. Kafatos, 'Notes from ERC Scientific Council Retreat', Informal Minutes (Les Treilles: ERC ScC, 16 October 2008), 10[TK].

97 '37th Scientific Council Meeting, Limassol, 4.–5.10.2012', Annotated Minutes (Brussels: ERC ScC Plenary #38, 4 December 2012), 5.

7 WIDE-RANGING EFFECTS

1 'Nobel Prize in Physics Goes to European Research Council Grantee Prof. Konstantin Novoselov', Press Release (Brussels: ERC ScC, 5 October 2010).

2 K. S. Novoselov et al., 'Electric Field Effect in Atomically Thin Carbon Films', *Science* 306, no. 5696 (22 October 2004): 666–9.

3 K. S. Novoselov, 'Physics and Applications of Graphene', Project Abstract (Cordis: Project Database Entry 88458, 1 December 2008).

4 'Nobel Prize Novoselov.'

5 Ibid.; Geoghegan-Quinn's predecessor once stated: 'I believe the ERC will successfully provide Europe with the world-leading capabilities in frontier research that are necessary to meet the global competitiveness challenge. And it will perhaps also help in increasing the number European scientists winning Nobel Prizes, be they European by nationality or their choice of place for doing research.' Janez Potočnik, 'Speaking Points London School of Economics', Unpublished Document (London, 25 April 2006), 6 [EC-AA].

6 'Nobel Prize Novoselov'; the Scientific Council dutifully continued to report grantees who became Nobel Laureates, cf. 'Nobel Prize in Physics 2012 Goes to European Research Council Grantee', Press Release (ERC ScC, 9 October 2012) [TK].

7 Karl Ulrich Mayer, 'From Max Weber's "Science as a Vocation (1917)" to "Horizon 2020"', Max Weber Lecture Series MWP – 2013/06 (Florence: European University Institute, 19 June 2013): 11.

8 'The ERC grants are addressed to individual researchers, but over time, they will also collectively illuminate the performance of individual countries, regions, and institutions.' Cf. Manolis Antonoyiannakis, Jens Hemmelskamp, and Fotis C. Kafatos, 'The European Research Council Takes Flight', *Cell* 136, no. 5 (2009): 807.

9 *Ideas Specific Programme*, 259–260.

10 Alexis Michel Mugabushaka, 'ERC Monitoring and Evaluation Strategy (part 1) and Roadmap of 2009 M & E Activities (part 2)', (Brussels: ERC ScC Plenary #21, 30 June 2009), 5; a previous version had been published in 2008, which distinguished four 'categories of ERC effects and impacts': 'ERC Science management & organisation', correlating to 'performance development'; 'Research themes & scientific output'; 'Researchers & host institutions', and 'Policy & structures', correlating with 'policy impact'; cf. *ERC Work Programme 2008*, 15–6.

11 According to the legal basis of the Executive Agencies, each had to 'submit

an annual activity report together with financial and management information.' *Executive Agencies*, Art. 9(7).

12 *Ideas Specific Programme*, 259–260.

13 Mugabushaka, 'Monitoring and Evaluation Strategy', 9.

14 'ERC Scientific Council Working Group on Key Performance Indicators, 2nd Meeting, Utrecht, 21 October 2013', Minutes (Brussels: ERC ScC Plenary #43, 3 December 2013), 6.

15 The original index is part of the bi-annual NSF Science and Engineering Indicators, based on the Thomson Reuters Science Citation Index (CSI) database. First the total of articles in a given year is assigned to world regions. Next, the expected number of articles in the 99th percentile of articles (according to citation rate) can be determined for each region. The resulting threshold is held against the actual share of top 1% cited articles per region; in other words, the index in the table represents the deviation from this threshold. See 'Science and Engineering Indicators 2012' (Arlington VA: NSF, 2012), Appendix 5–45, and 'Science and Engineering Indicators 2014' (Arlington VA: NSF, 2014), Appendix 5–57. Given the fact that the ERC had available to it only a tiny fraction of the overall R&D budget spent in Europe, its use of the index was intended to show its outstanding performance within the European landscape, and not in the global setting.

16 *ERC Work Programme 2008*, 15–6.

17 Thomas Reiss et al., 'ERACEP – Emerging Research Areas and Their Coverage by ERC-Supported Projects', Final Report (Brussels: ERCEA, April 2013), 4.

18 Katy Whitelegg et al., 'DBF Development and Verification of a Bibliometric Model for the Identification of Frontier Research', Synthesis Report (Brussels: ERCEA, February 2013), i.

19 Among others, cf. Wolfgang Glänzel and Bart Thijs, 'Using "core Documents" for Detecting and Labelling New Emerging Topics', *Scientometrics* 91, no. 2 (2011): 399–416; Wolfgang Glänzel, 'The Role of Core Documents in Bibliometric Network Analysis and Their Relation with H-Type Indices', *Scientometrics* 93, no. 1 (2012): 113–23; Marianne Hörlesberger et al., 'A Concept for Inferring 'frontier Research' in Grant Proposals', *Scientometrics* 97, no. 2 (2013): 129–48; Jörg Neufeld, Nathalie Huber, and Antje Wegner, 'Peer Review-Based Selection Decisions in Individual Research Funding, Applicants' Publication Strategies and Performance: The Case of the ERC Starting Grants', *Research Evaluation* 22, no. 4 (2013): 237–47; Thomas Scherngell et al., 'Initial Comparative Analysis of Model and Peer Review Process for ERC Starting Grant Proposals', *Research Evaluation* 22, no. 4 (2013): 248–57.

20 'ERC Monitoring and Evaluation Strategy. ERC CSA Support Projects 2008 and 2009', Draft (Brussels: ERCEA, 21 May 2010), 3 [TK]; similarly, Scientific Council member Carlos Duarte was quoted saying that the projects 'have not delivered according to expectations and access to sensitive information is difficult if not impossible' [for them], quoted in: 'Minutes 2nd KPI Working Group Meeting.'

21 Whitelegg et al., 'DBF', iv; to be fair, the sentence continued that scientometrics' 'potential for implementation within funding agencies was found relevant for exploring further.'

22 Maria Nedeva et al., 'Understanding and Assessing the Impact and Outcomes

of the ERC and Its Funding Schemes (EURECIA)', Final Report (Brussels: ERCEA, May 2012), 3.

23 Nathalie Huber, Antje Wegner, and Jörg Neufeld, 'MERCI (Monitoring European Ressearch Council's Implementation of Excellence): Evaluation Report on the Impact of the ERC Starting Grant Programme', IFQ Working Paper (Berlin: Institut für Forschungsinformation und Qualitätssicherung, December 2015), 166–9.

24 Luukkonen, 'Peer Review'; Nedeva, 'Between the Global and the National'; Neufeld, Huber, and Wegner, 'Peer Review-Based Selection Decisions'; Terttu Luukkonen, 'The European Research Council and the European Research Funding Landscape', *Science and Public Policy* 41, no. 1 (2014): 29–43.

25 Maria Nedeva and Michael Stampfer, 'From "Science in Europe" to "European Science"' *Science* 336, no. 6084 (25 May 2012): 983.

26 Grit Laudel and Jochen Gläser, 'Beyond Breakthrough Research: Epistemic Properties of Research and Their Consequences for Research Funding', *Research Policy* 43, no. 7 (2014): 1215.

27 Nedeva et al., 'EURECIA', 120.

28 Veugelers would soon also be appointed member of RISE (standing for Research, Innovation and Science Policy Experts), which was to advise the European Commission on the Innovation Union. In this capacity, she would write a policy brief on 'European public funding for research in the era of fiscal consolidation', cf. Reinhilde Veugelers, 'Is Europe Saving Away Its Future? European Public Funding for Research in the Era of Fiscal Consolidation', RISE Policy Brief (Brussels: European Commission, December 2014).

29 Reinhilde Veugelers, 'ERC-ScC Working Group on Key Performance Indicators First Meeting' (Vienna: ERC ScC KPI Meeting #1, 3 June 2013), 3.

30 'Minutes 2nd KPI Working Group Meeting', 2–3.

31 Based on a non-representative count of 'innovation' in the title of European Commission Communications between 2000 and 2013, the term appeared ten times until 2004, fourteen times between 2005 and 2007, seven times between 2008 and 2010, and seventeen times in 2011 alone; in 2012, it appeared six times, and in 2013 five times.

32 'Lisbon Strategy Evaluation Document', (Brussels: European Commission, 2 February 2010), 10 [TK].

33 Sapir et al., 'Growing Europe'; Wim Kok, 'Facing the Challenge. The Lisbon Strategy for Growth and Employment'(Luxembourg: European Commission, 1 November 2004).

34 Paul Zagamé, 'The Costs of a Non-Innovative Europe: What Can We Learn and What Can We Expect from the Simulation Works', n.n., 12.

35 *Europe 2020 A Strategy for Smart, Sustainable and Inclusive Growth* (European Commission: COM/2010/2020 [O.J. C 26/12], 3 March 2010); cf. also *Europe 2020 Flagship Initiative Innovation Union* (European Commission: COM/2010/546 [O.J. C 121/53], 6 October 2010).

36 Máire Geoghegan-Quinn, 'Towards an "I-Conomy" – Commissioner Máire Geoghegan-Quinn Delivers the 2010 Guglielmo Marconi Lecture at the Lisbon Council's Innovation Summit', Transcript (Rapid: SPEECH/10/68, 5 October 2010).

37 *Europe 2020*, 32.

38 *Green Paper: From Challenges to Opportunities: Towards a Common Strategic Framework for EU Research and Innovation Funding* (European Commission: COM/2011/48, 9 February 2011): 6.
39 Godin, *Making*, 63–4.
40 'Contribution of the ERC Scientific Council to the Consultation on the Common Strategic Framework for EU Research and Innovation Funding' Report (Brussels: ERC ScC, 15 May 2011), 7 [TK].
41 Ibid., 8.
42 *ERC Work Programme 2012*, 40.
43 Helga Nowotny, 'Innovation and Frontier Research', in *Innovation How Europe Can Take off*, ed. Simon Tilford and Philip Whyte (London: Centre for European Reform, 2011), 14.
44 Ibid., 15.
45 'Contribution', 7.
46 The Commission had come up with three suggestions to give the new edition a more notable name: 'imagine 2020', 'discover 2020', and 'horizon 2020'; cf. Máire Geoghegan-Quinn, 'The Future of EU-Funded Research and Innovation Programmes: An Emerging Consensus . . . and a New Name', Transcript (Rapid: SPEECH/11/432, 10 June 2011). Not surprisingly, the only noun won.
47 *Regulation (EU) Establishing Horizon 2020 the Framework Programme for Research and Innovation (2014–2020) and Repealing Decision No. 1982/2006/EC* (European Parliament and Council of the European Union: EU/2013/1291 [O.J. L 347/104], 11 December 2013).
48 In a later publication, which can also be read as a personal reflection of the experiences during her time with the ERC, Nowotny would state that 'politicians wish to be seen as innovating Europe out of the present crisis. The image conveys that all challenges can be met through innovation, however vague the concept remains. But it is doubtful that the future is only a scientific or technological challenge.' *The Cunning of Uncertainty* (Cambridge: Polity, 2016), 29.
49 Alain Peyraube, 'Report on the World Economic Forum Summer Davos in Asia – Annual Meeting of the New Champions 2012', (Brussels: ERC ScC, 1 October 2012) [TK].
50 'Secure the EU Research Budget – for a Future-Oriented Europe!', No-Cuts-on-Research.eu, accessed 15 September 2015, http://www.no-cuts-on-research.eu.
51 *Conclusions (multiannual Financial Framework)* (European Council: EUCO 37/13, 7–8 February 2013): 7: 'Given their particular contribution to the objectives of the Europe 2020 Strategy, the funding for Horizon 2020 and ERASMUS for all programmes will represent a real growth compared to 2013 level.'
52 'ERC Operations and Realization of Objectives 2013', 6.
53 Andrea Bonaccorsi, 'Ex-Post Evaluation of the Seventh Framework Programme. Support Paper to the High Level Expert Group. Ideas Specific Programme Analytical Evaluation' (Brussels: European Commission, March 2015), 30.
54 Louise O. Fresco et al., 'Commitment and Coherence. Ex-Post-Evaluation of the 7th EU Framework Programme (2007–2013)' (Brussels: European Commission, November 2015), 44.

55 'Comments by the European Research Council Scientific Council on the Ex-Post Report by the High Level Expert Group on the 7th Framework Programme for Research and Technological Development (FP7)' (Brussels: ERC ScC, 14 March 2016), 1.
56 Huber, Wegner, and Neufeld, 'MERCI', 130.
57 As the report on the previous edition of the Framework Programme had stated, 'the real impact on attractiveness and mobility [of ERC grants, TK] is low since most of the grantees were already in their institutions when they made their proposal.' Cf. Fresco et al., 'Ex-Post Evaluation Ideas Programme', 42.
58 Nedeva et al., 'EURECIA', 102.
59 Antonoyiannakis and Kafatos, 'ERC', 515.
60 Winnacker, 'On Excellence through Competition', 126.
61 'Commission to Cut Further Red Tape in Research Funding Procedures Questions and Answers', Press Release (Rapid: MEMO/10/156, 27 April 2010), 2.
62 Ironically, some academic institutions were against introducing this measure to the other programmes in the FP format, 'arguing that the flat rate could unfairly decrease their funding levels'; cf. Tania Rabesandratana 'Auditors Slam Red Tape at E.U. Science Funder', *Science Insider* (7 June 2013)
63 Nedeva et al., 'EURECIA', 110.
64 Sjoerd Hardeman, Vincent Van Roy, and Daniel Vertesy, 'An Analysis of National Research Systems (I): A Composite Indicator for Scientific and Technological Research Excellence', JRC Scientific and Policy Report (Brussels: Joint Research Centre Institute for the Protection and Security of the Citizen, November 2013).
65 Fresco et al., 'Ex-Post Evaluation Ideas Programme', 44; as the report further detailed: '12 organizations received more than 2 billion euros (26% of the FP7-IDEAS). 60% of the grants are attributed to the first 100 organizations. More than 90% are attributed to less than 600 organizations. The first 11 countries received more than 90% of the grants.'
66 Antonoyiannakis, Hemmelskamp, and Kafatos, 'ERC Takes Flight', 808.
67 Benjamin Turner, 'What Is Wrong with the Distribution of ERC Grants?' (Brussels: ERC ScC Plenary #37, 12 September 2012).
68 Fresco et al., 'Ex-Post Evaluation Ideas Programme: 44.'
69 'ERC Funding Activities 2007–13. Key Facts, Patterns and Trends' (Brussels: ERCEA, June 2015); 'Science behind the Projects. Research Funded by the European Research Council (2007–2013)' (Brussels: ERCEA, 13 October 2015).
70 'Funding Activities', 45; 39; 27.
71 Ibid., 35.
72 'Science behind the Projects', 26.
73 Elisabeth Pain, 'What Kinds of Science Do ERC Grantees Do?', *Science*, 3 November 2015.
74 'Science behind the Projects', 37.
75 'In view of [the activities'] science-driven nature and largely "bottom-up", investigator-driven funding arrangements, the European scientific community will play a strong role in determining the avenues of research followed under Horizon 2020.' Cf. *Horizon 2020*, 123.

8 SUMMARY

1 Vīķe-Freiberga et al., 'Towards', 24.
2 This paragraph is based on, and basically recounts, my analysis in König, 'Mission Accomplished?'
3 Winnacker, *Aufbruch*.
4 'Informal Minutes 17th Meeting, 10.–11.9.2008', 1.
5 'I am reaching my limits', as he declared as early as autumn 2008; cf. ibid., 8.
6 Potočnik, 'Speaking Points to Scientific Council.'
7 *Horizon 2020*, 123.
8 Gary Marks, 'Europe and Its Empires: From Rome to the European Union', *Journal of Common Market Studies* 50, no. 1 (2012): 7.
9 Mike Galsworthy, 'Brexit and Science: Let's Not Make the Same Mistake as the Swiss', *Guardian*, 17 June 2015; Ian Sample, 'Brexit Could Cost UK Science Millions in Lost Research Funding, Peers Warn', *Guardian*, 19 April 2016; Daniel Hook and Martin Szomszor, 'Examining Implications of Brexit for the UK Research Base. An Analysis of the UK's Competitive Research Funding', (London: Digital Science, May 2016).

9 POSTSCRIPT

1 Michèle Lamont, *How Professors Think: Inside the Curious World of Academic Judgment* (Cambridge, MA: Harvard University Press, 2009); Michèle Lamont and Katri Huutoniemi, 'Comparing Customary Rules of Fairness: Evaluative Practices in Various Types of Peer Review Panels', in *Social Knowledge in the Making*, ed. Charles Camic, Neil Gross, and Michèle Lamont (Chicago: University of Chicago Press, 2011), 209–32; Katri Huutoniemi, 'Communicating and Compromising on Disciplinary Expertise in the Peer Review of Research Proposals', *Social Studies of Science* 42, no. 6 (2012): 897–921.

INDEX

Note: page numbers in italics denote figures or tables